SpringerBriefs in Astronomy

SpringerBriefs in Astronomy are a series of slim high-quality publications encompassing the entire spectrum of Astronomy, Astrophysics, Astrophysical Cosmology, Planetary and Space Science, Astrobiology as well as History of Astronomy. Manuscripts for SpringerBriefs in Astronomy will be evaluated by Springer and by members of the Editorial Board. Proposals and other communication should be sent to your Publishing Editors at Springer.

Featuring compact volumes of 50 to 125 pages (approximately 20,000-45,000 words), Briefs are shorter than a conventional book but longer than a journal article. Thus Briefs serve as timely, concise tools for students, researchers, and professionals.

Typical texts for publication might include:

- A snapshot review of the current state of a hot or emerging field
- A concise introduction to core concepts that students must understand in order to make independent contributions
- An extended research report giving more details and discussion than is possible in a conventional journal article
- A manual describing underlying principles and best practices for an experimental technique
- An essay exploring new ideas within astronomy and related areas, or broader topics such as science and society

Briefs allow authors to present their ideas and readers to absorb them with minimal time investment.

Briefs will be published as part of Springer's eBook collection, with millions of readers worldwide. In addition, they will be available, just like other books, for individual print and electronic purchase.

Briefs are characterized by fast, global electronic dissemination, straightforward publishing agreements, easy-to-use manuscript preparation and formatting guidelines, and expedited production schedules. We aim for publication 8-12 weeks after acceptance.

More information about this series at https://link.springer.com/bookseries/10090

Debanjan Bose · Subhendu Rakshit

High Energy Astrophysical Neutrinos

Springer

Debanjan Bose
Department of Astrophysics
and Cosmology
S. N. Bose National Centre for Basic
Sciences
Kolkata, West Bengal, India

Subhendu Rakshit
Department of Physics
Indian Institute of Technology Indore
Indore, Madhya Pradesh, India

ISSN 2191-9100   ISSN 2191-9119 (electronic)
SpringerBriefs in Astronomy
ISBN 978-3-030-91257-4   ISBN 978-3-030-91258-1 (eBook)
https://doi.org/10.1007/978-3-030-91258-1

This Springer imprint is published by the registered company Springer Nature Switzerland AG
The registered company address is: Gewerbestrasse 11, 6330 Cham, Switzerland

Preface

Neutrino astronomy is an emerging research field encompassing domains of astronomy, astrophysics and particle physics. At very high energies, neutrinos are the only means to study our Universe as it is opaque to photons. The field met its first success when solar neutrinos were observed and then neutrinos from the supernova 1987A. Since then, neutrino astronomy has made tremendous progress over the last few decades, particularly with the IceCube detector at the South Pole. The era of high-energy neutrino astronomy started with the observation of astrophysical neutrinos with IceCube. Various other experiments are also coming up after this initial success.

While quite a few reviews have thus far been published on this topic, it is now pertinent to have a consolidated reading for students at the graduate level who have some basic exposure in astrophysics, particle physics and quantum field theory. The idea is to set the stage for them to start doing research in the field of neutrino astronomy. We present a brief review of topics related to the high-energy astrophysical neutrinos and other messengers and summarise ideas that may lead to future research on this area. The theoretical aspects and related experiments are both highlighted.

We have learned about the physics related to IceCube from the excellent reviews and lectures by Francis Halzen over the years, and this will be largely reflected in this exposition.

We are indebted to B. Ananthanarayan, IISc, Bangalore, for mobilising us to take up the assignment of writing this book, his guidance and continuous encouragement. Subhendu Rakshit (SR) is thankful to Ewald Reya, who initiated SR into this field of research. SR also benefited while collaborating with Siddhartha Karmakar and Sujata Pandey.

Debanjan Bose acknowledges Science and Engineering Research Board (SERB)—Department of Science and Technology, Government of India, for Ramanujan Fellowship—SB/S2/RJN-038/2017. This work is also supported by SERB grants received by SR: MTR/2019/000997 and CRG/2019/002354.

Kolkata, India   Debanjan Bose
Indore, India   Subhendu Rakshit
October 2021

Contents

Acronyms

AGN	Active Galactic Nuclei
BSM	Beyond the Standard Model of Particle Physics
CBR	Cosmic Background Radiation
CC	Charged Current
CMB	Cosmic Microwave Background
CR	Cosmic Ray
DIS	Deep Inelastic Scattering
DOM	Digital Optical Module
EBL	Extragalactic Background Light
GR	Glashow Resonance
GRB	Gamma Ray Burst
GZK	Greisen–Zatsepin–Kuzmin
HERA	Hadron-Electron Ring Accelerator
HESE	High-Energy Starting Events
IGM	Intergalactic Medium
LHC	Large Hadron Collider
MBH	Microscopic Black Hole
NC	Neutral Current
PDF	Parton Distribution Function
SBG	Star Bursts Galaxy
SNR	Supernova Remnant
UHE	Ultra-High Energy
VHE	Very-High Energy

Chapter 1
Introduction

In the standard model of particle physics (SM), amongst the elementary particles, neutrinos are the only fermions that do not carry any electric or colour charge. Apart from gravitational interactions, they take part only in the weak interactions. As neutrinos do not take part in strong and electromagnetic interactions, it can pass through matter in the astrophysical objects almost unhindered. Thus neutrinos can act as messengers that can carry information about the dynamics inside the astrophysical objects from the farthest corners of the Universe. As a result, neutrinos are regarded as reliable messengers for astronomy.

Photons are traditionally used as messengers in astronomy. The reason being, like neutrinos they are also charge neutral. So a photon does not get deviated by interstellar magnetic fields. It is also easy to produce and detect photons through electromagnetic interactions. That is why photon-based astronomy works for a vast range of wavelengths, starting from radio waves to gamma rays. But such astronomy comes with its limitations. As photons couple to charged particles electromagnetically, it can be easily absorbed by dust and other matter particles, restricting its penetrability. High energy gamma rays can get absorbed by the background photons. So for extragalactic astronomy, photons are not useful messengers beyond a few tens of TeVs. Neutrinos do not suffer from such limitations.

We have observed cosmic rays (CR) up to an energy $\sim 10^{11}$ GeV. The composition of CR vary from protons to heavier nuclei, such as iron. But charged particles lose their directionality as they travel through space. So only those with extreme energies can be used for astronomy, as they can point back to the sources.

The observation of ultra-high energy CR immediately suggest associated production of photons and neutrinos. As high energy photons cannot reach us, observation of high energy neutrinos is an important tool to understand various acceleration mechanisms that lead to the extremely energetic CRs. With the observation of several high energy neutrino events at IceCube, a new era of astronomy has begun. As we will point out, as these neutrinos interact with the nucleons at a centre-of-mass

© The Author(s), under exclusive license to Springer Nature Switzerland AG 2021   1
D. Bose and S. Rakshit, *High Energy Astrophysical Neutrinos*,
SpringerBriefs in Astronomy, https://doi.org/10.1007/978-3-030-91258-1_1

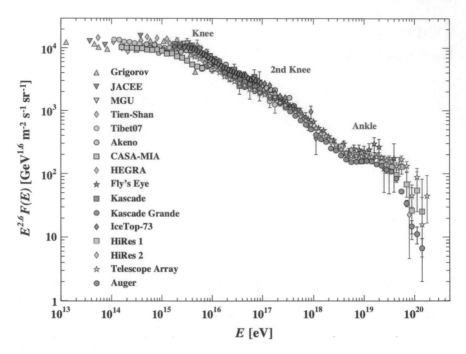

Fig. 1.1 The all-particle spectrum as a function of E (energy-per-nucleus) from air shower measurements. Reproduced with permission from Ref. [1]

energy that can exceed the same for the Large Hadron Collider (LHC) at CERN, one can explore new physics options beyond SM with the observations of high energy neutrinos.

1.1 Cosmic Rays

The observed cosmic ray spectrum exhibits a typical power law spectrum $\propto E^{-\alpha}$. Above a PeV, the spectrum gets steeper, so that the spectral index α changes approximately from 2.7 to 3. Often referred to as 'knee', this feature is believed to be associated with the maximum energy up to which the galaxy is capable of accelerating CR, given the size of our galaxy and the existing magnetic field. Around 100 PeV, α changes again from 3 to 3.3, which is known as the 'second knee'. Above 5 EeV, the spectrum flattens, shifting α from 3.3 to about 2.6, referred to as 'ankle'. Above 40 EeV, the flux shows a drastic fall, known as GZK cut-off, that follows from the interaction of CR with the background photons (Fig. 1.1).

At lower energies ~100 GeV, one expects to receive on an average one CR particle per square meter per second. Around a PeV this rate decreases to one particle per square meter per year. Above an EeV, this rate further drops down to one particle/km^2/year. In detectors like Auger, one can do astronomy with these particles for energies more than 10^9 GeV, as charged particles with lower energies deflect in the interstellar magnetic field, and fails us to point back towards its source. The air shower detection facilities designed to look for such particles can span over a few thousand sqkm.

1.2 How Cosmic Rays Get Accelerated?

The power law nature of cosmic ray spectrum suggests non-thermal methods to boost CRs to higher energies. Electric fields can accelerate charged particles. However, to boost the CRs to such high energies, one needs to sustain such a potential difference over a long distance. The presence of the charged plasma in these astrophysical objects is likely to prohibit such a configuration, making such an option rather unlikely. Such plasmas can carry with them large magnetic fields, that can vary over both space and time. Charged particles can get non-thermally accelerated in such configurations interacting with such plasma clouds that can act as magnetic reflectors [2].

The clouds are massive compared to the charged particles to get accelerated and move at a speed V, much less compared to the speed of these particles. For head-on collisions the particles gain energy, whereas for follow-on collisions it goes otherwise. Such interactions may happen repeatedly for a given particle. It can be shown that the head-on collisions are more likely to happen than the follow-on ones. This results in a net increase in the energy pumped in to the particle causing an acceleration. The average energy gain per collision turns out to be proportional to V^2. Originally proposed [3] by E. Fermi, this method of non-thermal acceleration of particles is known as 'second-order Fermi acceleration'.

The downside of the above-mentioned mode of particle acceleration can be understood as follows. The average gain in energy per collision $\langle \Delta E/E \rangle$ is proportional to $(V/c)^2$. In practice, $V \ll c$ and due to this the particles take considerable amount of time to gain sufficient energy. Moreover, as accelerated particles lose energy through emitting electromagnetic radiation, the rate of net energy gain reduces, making the process inefficient. Moreover, this process does not yield to a unique power law flux, where as the observations suggest an index $\sim$2–3.

The shortcomings of the second-order Fermi acceleration can be traced back to the fact that the particles can also lose energy in the follow-on collisions. So if the particle can somehow gain energy in each collision, such a bottleneck can be evaded. This is achieved [4–7] in the 'first-order Fermi acceleration', when charged particles come into the contact of shocks. Such shocks can be present in astrophysical plasma, for example, in the supernova remnants or in the jets of the quasars. Particles can be accelerated up to $\sim$100 TeV in supernova remnants.

In contrary to the second-order Fermi acceleration, here $\langle \Delta E/E \rangle \propto V/c$, justifying the nomenclature 'first-order'. This leads to a fixed power law spectrum $N(E) \propto E^{-2}$, under some simplified assumptions. For example, here the particles are considered to be relativistic, with no back-reaction of the particles onto the shock. The shocks are also considered non-relativistic. Depending on the typical astrophysical objects one is interested in, such assumptions may not be realised very strictly and one can reproduce the observed power law.

1.2.1 Hillas Criterion

The charged particles get accelerated by the transient electric fields and confined by the magnetic fields associated with the plasma. As we have seen cosmic rays as energetic as 10^{20} eV, it is important to find out which astrophysical objects can accelerate protons or iron nuclei to such energies. To achieve this, the acceleration sites should be large enough to confine these highly energetic particles, or in other words, the Larmor radius r_L should be smaller than the size of these astrophysical objects.

$$r_L = \frac{p}{ZeB} = \frac{E}{ZeBc} \simeq 1.1 \frac{1}{Z} \left(\frac{B}{\mu G} \right)^{-1} \frac{E}{\text{EeV}} \text{ kpc} \tag{1.1}$$

where Z stands for the charge of the particle in units of the charge of an electron. Thus, the size of these objects can be used to estimate an upper bound on the highest energy the particles can be accelerated to: $E_{max} \sim ZeBcL$. One can further take into account the speed β (in units of c) of magnetic inhomogeneities that act as scatterers. This dilutes the above estimate as

$$E_{max} \sim Z\beta eBcL = Z\beta \frac{B}{\mu G} \frac{L}{\text{kpc}} \text{ EeV} . \tag{1.2}$$

This implies that for an acceleration site of size L and magnetic field B, once the particle gets accelerated to E_{max}, it leaves the site and cannot be accelerated any further. Equation (1.2) is referred to as 'Hillas criterion' and can be visualised by putting various astrophysical objects in a plane spanned by L and B, first proposed by Hillas [8]. See Refs. [9, 10] for updated Hillas plots.

1.2.2 The Connection Between Cosmic Rays, Gamma Rays and Astrophysical Neutrino Spectra

At the source the accelerated protons can interact with background photons or matter producing neutrinos, photons and hadrons. The regions of acceleration and the site for such interactions should be separate for successful acceleration of particles to extreme energies. It is natural to expect that the spectra of cosmic rays, photons and neutrinos should be interrelated. Although one can do numerical simulations to establish such a connection, an approximate relationship can be derived [11–14] by taking some reasonable assumptions.

The protons interact with the photons producing pions, which eventually decay as follows:

$$p + \gamma \rightarrow \pi^0 + \text{anything}$$
$$\hookrightarrow \gamma\gamma \tag{1.3}$$

and

$$p + \gamma \rightarrow \pi^+ + \text{anything}$$
$$\hookrightarrow \mu^+ \nu_\mu$$
$$\hookrightarrow e^+ \nu_e \bar{\nu}_\mu . \tag{1.4}$$

The neutral pion decays immediately into a pair of photons via electromagnetic interactions. For the charged pions, their decay goes through weak interactions: $\pi^+ \rightarrow \mu^+ \nu_\mu$ and $\pi^- \rightarrow \mu^- \bar{\nu}_\mu$, where the muons again decay through the exchange of a W boson as: $\mu^+ \rightarrow e^+ \nu_e \bar{\nu}_\mu$ and $\mu^- \rightarrow e^- \bar{\nu}_e \nu_\mu$.

As mentioned in Sect. 1.3, in $p\gamma$ interactions charged and neutral pions are produced in the ratio $K_\pi = N_{\pi^\pm}/N_{\pi^0} = 1$. The inelasticity, defined as the fraction of the incoming proton energy released to the pion(s), is $\kappa_\pi = E_\pi/E_p \simeq 0.2$.

The secondaries e^+, ν_e, $\bar{\nu}_\mu$, and ν_μ produced from π^+ decay share the energy of the pion almost equally on an average. So, we will take $\kappa_\nu = E_\nu/E_\pi \simeq 0.25$. One can see that each neutrino carries a fraction $\kappa_\pi \kappa_\nu \simeq 0.05$ of the energy of the parent proton. Similarly, the photon inherits $\kappa_\pi \kappa_\gamma \simeq 0.1$ fraction of the proton energy.

The photons from π^0 decay will equally share its energy, so that, $\kappa_\gamma = E_\gamma/E_\pi = 0.5$.

In the hadronic collisions

$$p + p \rightarrow \pi + \text{anything}, \tag{1.5}$$

π^+, π^- and π^0 are produced in equal proportions, implying $K_\pi = 2$, with a corresponding inelasticity $\kappa_\pi \simeq 0.5$. Here, both pp and pn interactions are collectively termed as hadronic collisions.

As each of the charged pion gives rise to two muon and one electron flavoured neutrinos,

$$N_{\pi^\pm} = \frac{1}{2} N_{\nu_\mu} \implies \frac{dN_{\pi^\pm}}{dE_\pi} = \frac{\kappa_\nu}{2} \frac{dN_{\nu_\mu}}{dE_{\nu_\mu}}, \tag{1.6}$$

$$N_{\pi^\pm} = N_{\nu_e} \implies \frac{dN_{\pi^\pm}}{dE_\pi} = \kappa_\nu \frac{dN_{\nu_e}}{dE_{\nu_e}}. \tag{1.7}$$

As the present neutrino telescopes cannot distinguish between neutrinos and antineutrinos, here we refer to them as neutrinos only.

By the same token, for neutral pions decaying into two photons,

$$N_{\pi^0} = \frac{1}{2} N_\gamma \implies \frac{dN_{\pi^0}}{dE_\pi} = \frac{\kappa_\gamma}{2} \frac{dN_\gamma}{dE_\gamma}. \tag{1.8}$$

Now let us try to relate the flux of these secondary particles with the proton flux in a source. The differential proton flux $\frac{dN_p}{dE_p}$, in units of $\mathrm{GeV}^{-1}\mathrm{cm}^{-2}\mathrm{s}^{-1}$, is incident onto a target with optical depth τ. Each proton produces pions with multiplicity n_π.

From Eq. (1.45), the survival probability of a proton after travelling a distance τ through the target is given by $\exp(-\tau)$. Hence, the probability that the proton decays within an interval τ and $\tau + d\tau$ is given by $\exp(-\tau) - \exp(-\tau - d\tau) \simeq d\tau \exp(-\tau)$. Moreover, to simplify the calculations, let us assume that the produced n_π pions all have the same average energy $\langle E_\pi \rangle$. The rate of production of pions in the energy interval E_π to $E_\pi + dE_\pi$ is then denoted by $q_\pi(E_\pi)dE_\pi$, where

$$q_\pi(E_\pi) = \frac{dN_\pi}{dE_\pi dt} = \int dE_p \int_0^\tau d\tau' \exp(-\tau') \frac{dN_p}{dE_p dt} n_\pi \, \delta\left(E_\pi - \langle E_\pi \rangle\right). \tag{1.9}$$

Note that, here E_π is not an independent variable, but depends on E_p as $n_\pi \langle E_\pi \rangle = \kappa_\pi E_p$, in the case of a multi-pion production.

If the proton traverses a distance l in the target then the optical depth τ is related to the mean free path $\lambda(l) = 1/(n(l)\sigma)$ as

$$\tau = \exp\left(-\int_0^l \frac{dl'}{\lambda(l')}\right), \tag{1.10}$$

where $n(l)$ stands for the number density of the target and σ denotes the interaction cross-section of the proton with the target particles. We assume an isotropic target, so that $n(l) = nl$, and $\tau = nl\sigma$.

We make a further assumption that the cross-section is independent of energy, implying the optical depth is independent of E_p, so that in Eq. (1.9), the integrals on E_p and τ separate:

$$
\begin{aligned}
q_\pi(E_\pi) &= \int_0^\tau d\tau' \exp(-\tau') \int dE_p \frac{dN_p}{dE_p dt} n_\pi \, \delta\left(E_\pi - \langle E_\pi \rangle\right) \\
&= (1 - \exp(-\tau)) \int dE_p \frac{dN_p}{dE_p dt} n_\pi \, \delta\left(E_\pi - (\kappa_\pi/n_\pi)E_p\right) \\
&= (1 - \exp(-\tau)) \frac{n_\pi^2}{\kappa_\pi} \frac{dN_p}{dE_p dt}\bigg|_{E_p = n_\pi E_\pi/\kappa_\pi}.
\end{aligned} \tag{1.11}
$$

In astrophysical environments, the optical depths are usually small ($\tau \ll 1$), so that $(1 - \exp(-\tau)) \simeq \tau$. Hence, one can arrive at the following relation:

$$E_\pi^2 q_\pi(E_\pi) = f_\pi [E_p^2 q_p(E_p)]_{E_p=n_\pi E_\pi/\kappa_\pi} \tag{1.12}$$

where,

$$q_p(E_p) \equiv \frac{dN_p}{dE_p dt}(E_p), \tag{1.13}$$

and we have introduced a 'bolometric' factor $f_\pi = \tau \kappa_\pi \leq 1$.

For transparent sources, $f_\pi \ll 1$, so that the pion production is rather small. For more dense source, more pions are produced so the resulting flux of neutrinos and other secondaries also increase. For extremely dense source $f_\pi \simeq 1$, however, the pions can get absorbed in the target before they can decay, reducing the neutrino flux. In such scenarios, f_π should be replaced with $1 - \exp(-f_\pi)$.

For charged pions, Eq. (1.12) can be multiplied with the probability to produce $\pi^\pm$ out of all pions

$$P_{\pi^\pm} = \frac{N_{\pi^\pm}}{N_{\pi^\pm} + N_{\pi^0}} = \frac{K_\pi}{1 + K_\pi}, \tag{1.14}$$

so that

$$E_\pi^2 q_{\pi^\pm}(E_\pi) = f_\pi \frac{K_\pi}{1 + K_\pi}[E_p^2 q_p(E_p)]_{E_p=n_\pi E_\pi/\kappa_\pi}. \tag{1.15}$$

For neutral pions,

$$E_\pi^2 q_{\pi^0}(E_\pi) = f_\pi \frac{1}{1 + K_\pi}[E_p^2 q_p(E_p)]_{E_p=n_\pi E_\pi/\kappa_\pi}, \tag{1.16}$$

implying

$$q_{\pi^\pm}(E_\pi) = K_\pi q_{\pi^0}(E_\pi). \tag{1.17}$$

Then from Eqs. (1.6) and (1.8), one can obtain

$$q_{\nu_\mu}(E_\nu) = \frac{2}{\kappa_\nu}q_{\pi^\pm}(E_\pi)|_{E_\pi=E_\nu/\kappa_\nu} = \frac{2}{\kappa_\nu}K_\pi q_{\pi^0}(E_\pi)|_{E_\pi=E_\nu/\kappa_\nu} \tag{1.18}$$

$$q_{\nu_e}(E_\nu) = \frac{1}{\kappa_\nu}q_{\pi^\pm}(E_\pi)|_{E_\pi=E_\nu/\kappa_\nu} = \frac{1}{\kappa_\nu}K_\pi q_{\pi^0}(E_\pi)|_{E_\pi=E_\nu/\kappa_\nu} \tag{1.19}$$

$$q_\gamma(E_\gamma) = \frac{2}{\kappa_\gamma}q_{\pi^0}(E_\pi)|_{E_\pi=E_\nu/\kappa_\nu}. \tag{1.20}$$

As in the standard scenario, neutrino oscillations lead to 1:1:1 flavour ratios after traversing astrophysical distances,

$$\frac{1}{3}\sum_\alpha q_{\nu_\alpha}(E_\nu) = \frac{1}{3}\left(q_{\nu_e} + q_{\nu_\mu}\right) = \frac{1}{\kappa_\nu}q_{\pi^\pm} = \frac{K_\pi}{\kappa_\nu}q_{\pi^0} = \frac{K_\pi \kappa_\gamma}{2\kappa_\nu}q_\gamma(E_\gamma). \tag{1.21}$$

It implies,

$$\frac{1}{3}\sum_{\alpha} E_{\nu}^2 q_{\nu_\alpha}(E_\nu) = \kappa_\nu E_\pi^2 q_{\pi^\pm}(E_\pi) = \kappa_\nu f_\pi \frac{K_\pi}{1 + K_\pi}[E_p^2 q_p(E_p)]$$

$$= \frac{\kappa_\nu K_\pi}{2\kappa_\gamma} E_\gamma^2 q_\gamma(E_\gamma) = \frac{K_\pi}{4} E_\gamma^2 q_\gamma(E_\gamma) \qquad (1.22)$$

where, in the last step we have used numerical values of κ_ν and κ_γ.

Equation (1.22) describes for a point source how neutrino, photon and cosmic ray fluxes are related to one another. On earth an observer receives fluxes from a distribution of point sources, which can be estimated as follows.

The diffuse neutrino flux due to a point source distribution $\rho(r)$ is given by

$$\phi_\nu = \frac{1}{4\pi} \int d^3 r \rho(r) \frac{q_\nu(E_\nu)}{4\pi r^2} \qquad (1.23)$$

where, ϕ_ν carries the units $\text{GeV}^{-1} \text{cm}^{-2} \text{s}^{-1} \text{sr}^{-1}$. It can be simplified as

$$\phi_\nu = \frac{1}{4\pi} \int dr\, 4\pi r^2 \rho(r) \frac{q_\nu(E_\nu)}{4\pi r^2} = \frac{1}{4\pi} \int dr \rho(r) q_\nu(E_\nu). \qquad (1.24)$$

Although this works well for nearby sources, as the neutrinos can travel to us from faraway astrophysical objects, one need to modify the above equation including effects of the expansion of the Universe. We can transform the integration variable to redshift z noting that $dr = cdt = cdz(dt/dz)$ with $dz/dt = -(1 + z)H(z)$, so that,

$$\phi_\nu = \frac{c}{4\pi} \int \frac{dz}{H(z)} \rho(z)\, q_\nu((1 + z)E_\nu). \qquad (1.25)$$

In the standard model of cosmology,

$$H^2(z) = H_0^2[(1 + z)^3 \Omega_m + \Omega_\Lambda], \qquad (1.26)$$

with a cold matter component, $\Omega_m \simeq 0.3$, and the main dominance comes from the cosmological constant, $\Omega_\Lambda \simeq 0.7$. Here, $c/H_0 \simeq 4.4\,\text{Gpc}$.

Assuming the source flux follows a power law spectrum $q_{\nu_\alpha}(E_\nu) \propto E^{-\gamma}$,

$$\frac{1}{3}\sum_{\alpha} E_\nu^2 \phi_{\nu_\alpha}(E_\nu) = \frac{c}{4\pi} \frac{\xi_z}{H_0} \rho(0) \frac{1}{3}\sum_{\alpha} E_\nu^2 q_{\nu_\alpha}(E_\nu) \qquad (1.27)$$

where,

$$\xi_z = \int\limits_0^\infty dz \frac{(1 + z)^{-\gamma}}{\sqrt{(1 + z)^3 \Omega_m + \Omega_\Lambda}} \frac{\rho(z)}{\rho(0)}. \qquad (1.28)$$

contains the effects due to the expansion of the Universe.

In the local universe ($z < 2$), one can ignore the effects due to the expansion of the Universe, so that, $\rho(0) \equiv \rho_0$ is constant. In that case, $\xi_z \simeq 0.5$. However, if one considers star formation rate, $\xi_z \simeq 2.6$, where

$$\rho(z) = \begin{cases} (1+z)^3 \rho_0, & \text{for } z < 1.5 \\ (1+1.5)^3 \rho_0, & \text{for } 1.5 < z < 4. \end{cases} \tag{1.29}$$

In both cases, the spectral index $\gamma \simeq 2$.

Using Eq. (1.22) on the right hand side of Eq. (1.27) helps us relate the neutrino flux received at earth to the CR flux at the source.

$$\frac{1}{3} \sum_\alpha E_\nu^2 \phi_{\nu_\alpha}(E_\nu) = \frac{c}{4\pi} \frac{\xi_z}{H_0} \kappa_\nu f_\pi \frac{K_\pi}{1+K_\pi} [E_p^2 \mathcal{Q}_p(E_p)] \tag{1.30}$$

where the local emission rate density $\mathcal{Q}_p(E_p) \equiv \rho_0 q_p(E_p)$ can be estimated from the measured spectra as [15]

$$[E_p^2 \mathcal{Q}_p(E_p)]_{E_p=3\times10^{19}\,\text{eV}} \sim (1-2) \times 10^{44} \text{erg/Mpc}^3/\text{year}. \tag{1.31}$$

Note that the measurements suggest that the UHECRs are not only protons, but at high energies it receives some contribution from heavy nuclei. Spectral emission rates for nuclei with mass number A can be combined to write in terms of nucleon emission rate as $\mathcal{Q}_N(E_N) = \sum_A A^2 \mathcal{Q}_A(AE_N)$. It follows from $E_A/E_N = A = N_N/N_A$. Proton models yield similar results as other UHECR models with mixed composition so long as the spectral index $\gamma \simeq 2$. Auger measurements [16] indicate a combined nucleon-nuclei density $[E_N^2 \mathcal{Q}_N(E_N)]_{E_N=3\times10^{19}\,\text{eV}} \sim 2.2 \times 10^{43} \text{erg/Mpc}^3/\text{year}$ with $\gamma \simeq 2.04$.

For pp interaction ($K_\pi = 2$), one can then write,

$$\frac{1}{3} \sum_\alpha E_\nu^2 \phi_{\nu_\alpha}(E_\nu) = 3 \times 10^{-8} f_\pi \left(\frac{\xi_z}{2.6} \right)$$

$$\times \left(\frac{[E_p^2 \mathcal{Q}_p(E_p)]_{E_p=3\times10^{19}\,\text{eV}}}{10^{44} \text{erg/Mpc}^3/\text{yr}} \right) \text{GeV cm}^{-1} \text{s}^{-1} \text{sr}^{-1}. \tag{1.32}$$

In the calorimetric limit $f_\pi \to 1$, the above estimate is known as the Waxman-Bahcall limit [17, 18] on the flux of neutrinos as expected from the observation of UHECR spectrum. It is an upper limit in the sense that it is derived under the assumption that the source is thick, to produce the maximum neutrino flux. It is also assumed that the proton transfers all its energy to the pions inside the sources.

The connection between the UHECR flux and the observed neutrino flux is illustrated in Fig. 1.2 by the green lines. The solid green line represents an all-proton flux that is GZK-suppressed. It agrees with the observed CR flux mentioned in Eq. (1.31) and is aligned with the CR spectrum at the highest energies. The observed CR spec-

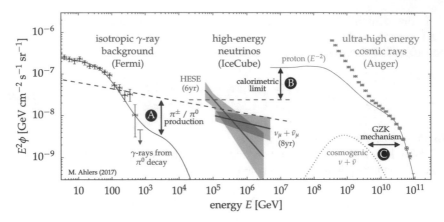

Fig. 1.2 Correlating the observed high energy neutrino spectrum with gamma-rays and cosmic rays. Reproduced with permission from Ref. [19]

trum is marked as green data points as measured by Auger. To explain the discrepancy at lower energies one needs additional CR sources. The green dashed line represents the diffuse neutrino flux as expected from Eq. (1.32). As neutrinos carry about 5% of the energy of the parent proton, the diffuse neutrino flux below 10 PeV can be explained from the lower part of the proton spectrum. The higher part of the spectrum after GZK suppression gives rise to the cosmogenic neutrinos, denoted by the green dotted line.

It is interesting to note that the above estimation of diffuse neutrino flux is close to the neutrino flux measurements at IceCube—the magenta line represents neutrino flux as measured from high energy starting events, and the red line stands for the neutrino flux as derived from the observation of muon track events alone. The corresponding bands indicate uncertainties at 1σ.

The connection between observed gamma ray flux by Fermi telescope and the observed neutrinos can also be established as illustrated in Fig. 1.2 by the blue lines. From Eq. (1.22), as one can relate neutrino and photon fluxes in a source, such a task looks feasible. Assuming that the cosmic ray spectrum follows a power law $\propto E^{-\gamma}$, one can see that photons and neutrinos also carry the same spectral index, but with different normalisations and energy scales are different as well.

The photons, as mentioned in Sect. 1.3, during propagation, interacts with the background photons leading to a cascade, that results in production of multiple photons in the GeV-TeV range. As Fermi can measure such photons, estimation of the corresponding neutrino flux is straightforward [20].

In Fig. 1.2, the source emissivity is normalised such that the solid blue line, representing the photon flux, does not exceed the isotropic photon background as seen by the Fermi and represented by the blue data points. The photon flux gets attenuated at energies more than 100 TeV due to the absorption in the background photons. The accompanying upper limit on neutrinos is shown by the blue-dashed line. The HESE neutrino data saturates this limit above $100\,\text{TeV}$.

The above-mentioned deliberations indicate that multimessenger studies with cosmic rays, neutrinos, and photons are extremely important tools that can help us understand particle production mechanisms at source and probe various proposed models in this regard.

1.3 GZK Cutoff and Cosmogenic Neutrinos

As cosmic rays pass through the Universe, they can lose energy. An important source is the expansion of the Universe. If a proton reaches the earth with an energy E_p from an object at a redshift z, taking into energy loss due to this expansion one can expect that the initial energy of the particle as $E_p(1+z)$. This implies that a particle would lose 50% of its energy before it is detected traversing a distance (range) $R \simeq 10^{28}$ cm $\simeq 3.3\,\text{Gpc}$.

Cosmic rays can interact with the cosmic background photons. Protons can interact with CMB photons as

$$p + \gamma_b \rightarrow p + e^+ + e^-. \tag{1.33}$$

The threshold energy for the proton to initiate such a process is $E_{p,th} = m_e^2/E_b \sim 10^{15}$ eV as CMB photons have a typical energy of $E_b \sim 10^{-3}$ eV. In each interaction, the proton would lose energy proportional to m_e/m_p. Such interactions would continue till the energy of the proton falls below the threshold. This helps us to define a range corresponding to this mode as

$$R = \frac{m_p}{2m_e} \frac{1}{\sigma_{pe^+e^-}n_b}. \tag{1.34}$$

For CMB, the number density $n_b \sim 400\,\text{cm}^{-3}$. The cross-section $\sigma_{pe^+e^-} \sim 10^{-27}\,\text{cm}^2$. Hence, $R \sim 10^{27}$cm $= 324\,\text{Mpc}$.

After the discovery of CMB, Greisen [21], Zatsepin and Kuzmin [22] realised that processes like

$$p + \gamma_b \rightarrow p + \pi^0$$
$$p + \gamma_b \rightarrow n + \pi^+ \tag{1.35}$$

can attenuate proton flux if its energy is above the threshold

$$E_{p,th} = \frac{m_\pi^2 + 2m_p m_\pi}{4E_b} \sim 10^{20}\text{eV}, \tag{1.36}$$

where we do not distinguish between proton and neutron masses. This threshold energy is known as Greisen-Zatsepin-Kuzmin(GZK) cut-off. One can easily check that in the proton rest frame this corresponds to the photon having an energy more than $m_\pi + m_\pi^2/(2m_p) \sim 150\,\text{MeV}$, which is little above the rest mass of pion $m_\pi \sim 140\,\text{MeV}$. In each interaction, the energy loss of the proton is proportional to m_π/m_p, i.e., it sheds off on an average 15–20% of its energy. For higher energy protons multi-pion production processes dominate, which may lead to about 50% energy loss of the proton. But the cross-section of single pion production process is about six times less than that for multi-pion processes.

The range is then given by

$$R = \frac{m_p}{m_\pi} \frac{1}{\sigma_{p\gamma} n_b} \sim 3 \times 10^{24}\text{cm} \simeq 10\,\text{Mpc}, \tag{1.37}$$

where $\sigma_{p\gamma} \sim 6 \times 10^{-28}\,\text{cm}^{-2}$.

In practice, both of the processes in Eq.(1.35) contribute, but the first one is twice as likely to happen than the second one. Moreover, it is the interaction with the higher energy photons in the tail of the CMB spectrum that decides the exact GZK cut-off around $6 \times 10^{19}\,\text{eV}$. Above this energy, the horizon for such cosmic ray sources reduce drastically – 90% of events with $E_p > 10^{20}\,\text{eV}$ come from distances $R < 100\,\text{Mpc}$.

The nuclei in the cosmic rays can also interact with the background photons leading to photo-disintegration to a lighter nuclei and typically one or two nucleons:

$$A + \gamma_b \rightarrow (A - 1) + N. \tag{1.38}$$

Above $10^{19}\,\text{eV}$ the CMB and the CIB photons interact with the nuclei predominantly via the 'giant dipole resonance' (GDR) [23], that leads to emission of one or two nucleons from the nuclei and α particles. Here, the actual threshold energies are dictated by the binding energy per nucleon $\sim 10\,\text{MeV}$. For all nuclei, GDR has the largest cross-section with thresholds between 10–20 MeV in the nucleus rest frame. Around 30 MeV the quasi-deuteron (QD) processes become comparable to GDR and dominates after that till 150 MeV—the photopion production threshold [24]. The threshold for flux suppression changes from $3 \times 10^{19}\,\text{eV}$ for He to $8 \times 10^{19}\,\text{eV}$ for Fe. Interaction lengths for Fe are similar to that of proton around GZK cut-off energy.

The pair production process(1.33) starts around $E_p = 2 \times 10^{18}\,\text{eV}$ and continues till $5 \times 10^{19}\,\text{eV}$, when the pion production(1.35) starts to dominate. Proton absorption due to this onset of pair production process was utilised in the 'dip model' to explain the existence of 'ankle' in the CR spectrum [25–27].

Pion production cross-section attains its maximum when a Δ^+ particle with rest mass of 1232 MeV is produced as an intermediate state, causing a resonance. Following isospin arguments, such reactions will yield pions in the ratio $N_{\pi^+} : N_{\pi^0} = 1:2$. Direct pion production can also driven by virtual meson exchanges, producing mostly

π^+, that contributes about 20% to the total $p\gamma$ cross-section. This implies that ultimately charged and neutral pions are produced in almost equal numbers in $p\gamma$ processes.

Neutral pions decay promptly as $\pi^0 \rightarrow \gamma + \gamma$. These photons interact with the CBR photons leading to pair production as $\gamma + \gamma_b \rightarrow e^+ + e^-$. These high energy electrons and positrons can in turn interact with background photons via inverse Compton scattering as $e + \gamma_b \rightarrow \gamma + e$, boosting up the energy of the photons. Processes like $\gamma + \gamma_b \rightarrow e^+ + e^- + e^+ + e^-$ and $e^\pm + \gamma_b \rightarrow e^\pm + e^+ + e^-$ also contribute [28]. Such interactions continue and lead to electromagnetic cascades. Part of the energy of the cascades go away from synchrotron radiation of the charged particles if the surrounding magnetic field is strong enough. The energy of the photons finally degrades to a level that they fail to produce e^+e^- pairs interacting with the diffuse optical photon background. After this the energy loss is mainly due to the expansion of the universe. Ultimately this leads to a pile-up of photons in the GeV-TeV range.

Charged pion decays via weak interactions as $\pi^+ \rightarrow \mu^+ + \nu_\mu$, with μ^+ subsequently decaying into $\mu^+ \rightarrow e^+ + \nu_e + \bar{\nu}_\mu$. The positron can lose its energy undergoing inverse Compton scattering with CRB and synchrotron radiation as mentioned before. The neutron decays as $n \rightarrow p + e^- + \bar{\nu}_e$ before it could get absorbed in the surrounding CMB radiation. The $\bar{\nu}_e$ carries only a tiny fraction $\sim 10^{-5}$ of the neutron energy. As the pion inherits about 20% of the energy of the proton, and that all the four particles from π^+ decay get almost equal share of pion energy, each of the three neutrinos $- \nu_\mu$, ν_e and $\bar{\nu}_\mu$ comes out with about 5% of the energy of the proton. These neutrinos, with energies predominantly in the EeV range, are known as 'cosmogenic' neutrinos [29], a byproduct of GZK effect.

Cosmogenic neutrinos are the most likely sources of astrophysical neutrino flux, as we have measured both ultra-high energy cosmic rays (UHECR) and CMB. However, we mention a few ingredients, that decide the flux of such neutrinos, and are not that well known.

• We are not certain about the chemical composition of UHECR. Are they protons, or heavy nuclei, or a composition of both? For heavy nuclei, higher the number of nucleons having energy greater than the pion production threshold, larger is the resulting cosmogenic neutrino flux.
• We are also not sure about the transition energy around which extragalactic component reigns over the galactic one. There are different propositions, and this gives rise to another source of uncertainty in the neutrino flux. In the aforesaid dip model, the transition occurs around $E \sim 3 \times 10^{16-17}$ eV, the second knee. The region from the second knee to the ankle is seen as a dip caused by pair production energy losses during propagation through the intergalactic medium.
• As the cosmogenic neutrinos are caused by the GZK effect, the question that usually arises is whether one has actually witnessed the GZK cut-off, or it is the manifestation of the maximum energy reach E_{max} of the cosmic accelerators. A doubt in this regard naturally arises as we have seen a few events beyond the

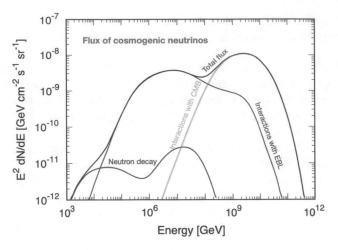

Fig. 1.3 Cosmogenic all-flavour neutrino flux considering pure proton CR composition, the dip transition model and star formation rate type source evolution. For the original plot see Ref. [32]

GZK cut-off, but yet to identify their sources. The cosmogenic neutrino spectra is sensitive to the value of E_{max}.

- The measurement of UHECR is limited by statistics. We are ignorant about the average acceleration spectrum of UHECR. A flatter spectrum obviously allow for more such particles to interact with the CMB.
- We are yet to identify the sources for UHECR, but they must evolve with redshift. This means their luminosity and spatial density should vary with redshift, although at present such cosmological evolution is not known for all prospective sources. Here, we need to keep in mind that although the UHECRs mainly can come from within a horizon of 100 Mpc or so, the neutrinos can travel uninterrupted from UHECR sources placed much beyond. Another critical component is the proton injection spectrum of these sources. All these contribute towards an uncertainty in the total source emissivity, reflecting the same in the expected neutrino flux.

In general, lighter compositions of UHECR, higher E_{max}, hard spectrum $\propto E^{-\gamma}$ with $\gamma \simeq 2$, and a strong redshift evolution of the sources lead to an increased flux for cosmogenic neutrinos. Strong evolution models also predict a large contribution to gamma rays. The extragalactic photon background in the GeV-TeV range, as indicated from the measurements of Fermi-LAT, can be used to set an upper limit on the energy density of cosmogenic neutrinos [30, 31].

In Fig. 1.3, Kotera et al. [32] had computed the cosmogenic neutrino flux assuming pure proton composition of cosmic rays. They used the star formation rate as obtained by Hopkins and Beacom [33] for star evolution where the source emissivity increases with redshift z as: $(1 + z)^{3.4}$ for $z < 1$, then $(1 + z)^{-0.26}$ for $1 \leq z < 4$ and $(1 + z)^{-7.8}$ for $z \geq 4$. For the galactic to extragalactic transition, the dip-model has been used. A proton injection spectrum $dN/dE \propto E^{-\alpha}$, with $\alpha = 2.5$ between energies 10^{16} eV and 3.2×10^{20} eV has been utilised. One can clearly see that the first hump in the flux around a PeV is caused by the interaction of cosmic rays with the infrared, optical and UV component of the EBL. Similar interaction with the CMB causes the second hump in the EeV range. Neutrons produced in the photo-hadronic dissociation processes decay as $n \rightarrow p + e^- + \bar{\nu}_e$. Neutrinos thus produced receive a tiny share of the proton energy as mentioned earlier. Dip-model leads to a large neutrino flux. The height of the PeV hump is mainly decided by the index α of the injection spectrum. The EeV hump depends moderately on α and the transition models. However, the major dependence comes from the chemical composition of UHECRs and E_{max}, the maximum acceleration energy.

Detection of cosmogenic neutrinos is an important task which the present neutrino telescopes are yet to achieve. IceCube has been designed to observe at least one such neutrino per year taking the most pessimistic flux into account. However, IceCube-Gen2 is expected settle the issue. Detection of these neutrinos will indirectly enhance our knowledge about the spectrum of UHECR and its composition. Getting an upper limit on the cosmogenic neutrino spectrum from observations would be important to identify prospective sources of UHECRs due to the dependence on the injection spectrum.

1.4 Photon Background

Our Universe is filled with electromagnetic radiation ranging from radio waves to gamma rays, the energy ranging over almost 20 decades. The microwave regime is dominated by cosmic microwave background (CMB) radiation. Extragalactic background light (EBL) stands for the integrated luminosity of all of the light emitted from the beginning of the universe till date. The light can be emitted from all kinds of sources—large objects like stars, galaxies etc., as well as from the small ones: Atoms, dust particles etc.

In Fig. 1.4, along the y-axis, νI_ν is plotted, which reflects how the energy density varies with the wavelength λ as I_ν is proportional to the energy density per unit frequency. At different wavelengths, the origin of such light can be different. One may note that in the literature, sometimes radio and microwave backgrounds are not included in the definition of EBL, and the entire background spectrum is termed as Cosmic background radiation (CBR). We will follow this convention in the rest of this book. Below we mention possible sources of CBR at different energy regimes [34].

- Cosmic radio background (CRB) [$\lambda > 30$ mm, $E < 4 \times 10^{-5}$ eV]: It is a combination of the following: Synchrotron radiation from charged particle passing through diffuse galactic and intergalactic magnetic fields, emission from AGN, H1 line emission, low energy tail of CMBR. Synchrotron radiation is proportional to B^2, where the magnetic field inside galaxies are small ($\sim 10^{-9}$ T), that explains why it is placed in the radio band.

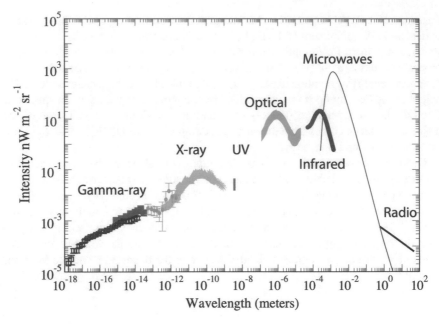

Fig. 1.4 Intensity νI_ν of extragalactic background light at different wavelengths as estimated from various measurements. The thickness of the curves are reflective of the relative uncertainties: Cosmic microwave background is known better than 1%, whereas, the optical background comes with large uncertainties. Wavelength range from 10 to 100 nm, that corresponds to UV, remains poorly explored. Reproduced from Ref. [35]

- Cosmic microwave background (CMB) [$\lambda : 0.3-30$ mm, $E : (0.04-4) \times 10^{-3}$ eV]:
  It is a relic from the early phase of the universe. The emission took place approximately 400,000 years after the Big Bang, as the universe cooled down, allowing protons to combine with electrons to form neutral atoms. This made the medium optically thin, so that it could not trap the photons any more. Although the radiation corresponds to that from a blackbody of temperature $\sim$3000 K at the epoch it got emitted, today it corresponds to the same at the temperature of 2.7255 ± 0.0006 K, due to the redshift $\sim$1100. CMB comes with the highest amplitude amongst all other components of CBR. The spectral radiance $dE/d\lambda$ peaks around a wavelength of 1.06 mm, with a corresponding frequency of 282 GHz. The energy of such photons turns out to be 1.2×10^{-3} eV. The average number density is $\sim$400/cm^3.
- Cosmic infrared background (CIB) [$\lambda : (3-300) \times 10^{-3}$ mm, $E : (4-400) \times 10^{-3}$ eV]:
  It is approximately half of the total energy density of the electromagnetic radiation emitted by the stars since the beginning of the universe. It primarily consists of emission (absorption and re-emission) from the dust particles, which got irradiated by the light emitted by the stars inside the galaxies. Early, most redshifted galaxies contribute to the long wavelength end of the spectrum. Hence, CIB is linked to the history of galaxy formation.

- Cosmic optical background (COB) [$\lambda : (0.3-3) \times 10^{-3}$ mm, $E : 0.4-4$ eV]:
  It consists of light directly coming from stars and hence, is linked with the history
  of cosmic star formation. Although such optical photons can be easily detected
  with ground based telescopes, the emission from the dust in the Milky Way acts as
  a foreground, thereby introducing considerable uncertainties in the measurements.
- Cosmic ultraviolet background (CUB) [$\lambda : (30-300) \times 10^{-6}$ mm, $E : 4-40$ eV]:
  The contributions come from the light emitted by young stars and interstellar neb-
  ulae, light scattered by dust particles, and perhaps emission from hot inter-cluster
  gas. This band remains rather poorly explored as the ground based telescopes
  are not suitable for such studies. Moreover, the efficiency of neutral hydrogen to
  absorb UV radiation renders interstellar medium opaque to these frequencies.
- Cosmic X-ray background (CXB) [$\lambda : (0.03-30) \times 10^{-6}$ mm, $E : 0.04-40$ keV]:
  Thermal bremsstrahlung X-ray photons from the accretion disks around AGNs are
  believed to be the major contributors in this frequency band. The gravitational pull
  that helps in the accretion, fuels such radiation. This is in contrast to the radiation in
  the lower frequency bands, which originates from the thermal and nuclear energy.
- Cosmic γ-ray background (CGB) [$\lambda < 3 \times 10^{-8}$ mm, $E > 40$ keV]:
  Contributions to γ-ray band comes mainly from the quasars/blazars and supernova explosions. In the first case, the ultra-relativistic charged particles in the jet
  can boost the surrounding photons at extreme energies via Compton scattering.
  Supernova blasts occur when a massive star loses its fuel, causing a core collapse, thereby generating a huge outburst of energy, a part of which gets emitted
  as $e^+ e^-$-pair. However, as we will see, such photons get converted into an $e^+ e^-$-pair
  interacting with the CIB photons, causing a cut-off around 300 GeV.

1.5 Photon Propagation

As photons propagate through galactic or intergalactic space, they can interact with
the photons that belong to CBR, producing dominantly an electron-positron pair.

$$\gamma + \gamma_b \rightarrow e^+ + e^-. \tag{1.39}$$

From the kinematics, using natural units, one can ascertain that for a given energy E
of the incident photon, the pair production happens if the energy of the background
photon E_b has an energy more than

$$E_{b,th}(E, \theta) = \frac{2m_e^2}{E(1 - \cos\theta)} \tag{1.40}$$

where, θ is the scattering angle and m_e stands for the electron mass. The cross-section
is given by [36, 37]

$$\sigma_{\gamma\gamma}(E, E_b, \theta) \simeq 1.25 \times 10^{-25} W(\beta) \, \text{cm}^2 \tag{1.41}$$

with,

$$W(\beta) = (1 - \beta^2) \left[2\beta(\beta^2 - 2) + (3 - \beta^4) \ln \frac{1 + \beta}{1 - \beta} \right]. \qquad (1.42)$$

β is the speed of e^+ and e^- in the CM frame, given by

$$\beta(E, E_b, \theta) = \left[1 - \frac{E_{b,th}}{E_b} \right]^{1/2}. \qquad (1.43)$$

$\sigma_{\gamma\gamma}$ is maximal for

$$E_b \simeq \left(\frac{900\,\text{GeV}}{E} \right) \text{eV}, \qquad (1.44)$$

assuming an isotropic background of photons [38].

In the above we have not explicitly shown the effect of the expanding universe. If the sources are placed at higher redshifts, then this can be easily taken into account noting that energy scales with a factor $1 + z$, where the z is the redshift of the source.

Equation (1.44) is quite useful in identifying which part of CBR can absorb photons of a given energy as mentioned below.

- For photons of energy $E < 10\,\text{GeV}$, the horizon is farther away than the Hubble radius, as the strength of EBL below UV is negligible compared to that for COB, CIB and CMB. This implies such photons can reach us almost uninterrupted due to lack of absorptions in the CBR.
- An incident photon of energy $E \sim 10\,\text{GeV}$ will suffer maximum absorption for $E_b \sim 90\,\text{eV}$ photons, that belong to the far UV band.
- $E \sim 10^5\,\text{GeV}$ similarly corresponds to $E_b \sim 10^{-2}\,\text{eV}$, which is in the far-infrared band.
- Similarly, incident photons with energy $10^5 \lesssim E \lesssim 10^{10}\,\text{GeV}$ get absorbed in the CMB background.
- Photons of energy greater than $10^{10}\,\text{GeV}$, interacts mostly with the CBR.

The photon absorption suggests that for a given energy of photons, there exists a horizon—from sources beyond that, the photons cannot reach us. A PeV energy photon is unlikely to reach us from the galactic center as it can interact with the CMB photons on its way.

The probability that a photon of observed energy E emitted from a source at a redshift z away from us will reach us is defined by the survival probability

$$P = \exp[-\tau(E, z)], \qquad (1.45)$$

where $\tau(E, z)$ is known as the optical depth. For sources not too far away, it can be expressed as [37] $\tau = D/\lambda_\gamma(E)$, where D is the distance of the source and $\lambda_\gamma(E)$ is the mean free path that depends on the energy of the photon. The aforesaid horizon now can be quantitatively defined by demanding $\tau = 1$. The energy dependence of λ_γ can be found in Refs. [37, 39]. For $E \sim 80\,\text{TeV}$, λ_γ is comparable to the distance

$3 - 5\,\mathrm{Mpc}$ of Centaurus A from us. As expected, for $1\,\mathrm{PeV}$ photons, $\lambda_\gamma \sim 8\,\mathrm{kpc}$—comparable to the distance of the galactic centre from the earth.

1.6 Neutrino Propagation

1.6.1 Neutrino Oscillations in Vacuum

When an electron undergoes a charged current interaction via the exchange of a W boson, an electron flavoured neutrino is produced. As we know today, there exists three such neutrino flavours: ν_α, where $\alpha = e, \mu, \tau$. However, the neutrinos propagate as mass eigenstates ν_i with $i = 1, 2, 3$, which are superpositions of the flavour eigenstates ν_α:

$$|\nu_i\rangle = \sum_\alpha U_{\alpha i}|\nu_\alpha\rangle. \tag{1.46}$$

U is a unitary mixing matrix, known as the Pontecorvo–Maki–Nakagawa–Sakata (PMNS) matrix, whose elements are determined from neutrino oscillation experiments.

ν_i are eigenstates of the Hamiltonian with energy eigenvalues (in natural units $\hbar = c = 1$)

$$E_i = \sqrt{|\mathbf{p}|^2 + m_i^2} \tag{1.47}$$

where $\mathbf{p}$ is three-momentum and m_i are the masses of these neutrinos. For relativistic neutrinos, the following approximation holds

$$E_i \simeq E + \frac{m_i^2}{2E}. \tag{1.48}$$

The mass eigenstates evolve with time as

$$|\nu_i(t)\rangle = e^{-iE_i t}|\nu_i(t = 0)\rangle. \tag{1.49}$$

Hence, the flavour eigenstate at time t is given by

$$|\nu_\alpha(t)\rangle = \sum_i U_{\alpha i}^\star e^{-iE_i t}|\nu_i(t = 0)\rangle = \sum_\beta \sum_i U_{\alpha i}^\star U_{\beta i} e^{-iE_i t}|\nu_\beta\rangle. \tag{1.50}$$

which reflects the underlying assumption that at $t = 0$, neutrino has a specific flavour α,

$$|\nu_\alpha(t = 0)\rangle \equiv |\nu_\alpha\rangle. \tag{1.51}$$

But after some time t, if the neutrino hits the detector, then the probability to detect a neutrino of flavour β is given by

$$P_{\alpha\beta} = |\langle \nu_\beta | \nu_\alpha(t) \rangle|^2 = \left| \sum_i U_{\alpha i}^\star U_{\beta i} e^{-iE_i t} \right|^2 = \sum_{i,j} U_{\alpha i}^\star U_{\beta i} U_{\alpha j} U_{\beta j}^\star e^{-i(E_i - E_j)t}.$$

(1.52)

Use of Eq. (1.48) then leads to the following expression for the neutrino oscillation probability from flavour α to β with $\alpha \neq \beta$

$$P_{\alpha\beta} = \sum_{i,j} U_{\alpha i}^\star U_{\beta i} U_{\alpha j} U_{\beta j}^\star \exp\left[-i \frac{\Delta m_{ij}^2}{2E} L \right],$$

(1.53)

with $\Delta m_{ij}^2 = m_i^2 - m_j^2$ ($m_i > m_j$) and we have replaced the time with the length of the path travelled L during time t.

Equation (1.53) can also be written as

$$P_{\alpha\beta} = \sum_i |U_{\alpha i}|^2 |U_{\beta i}|^2 + 2Re \sum_{i>j} U_{\alpha i}^\star U_{\beta i} U_{\alpha j} U_{\beta j}^\star \exp\left(-2\pi i \frac{L}{L_{\rm osc}} \right),$$

(1.54)

where,

$$L_{\rm osc} = \frac{4\pi E}{\Delta m_{ij}^2}.$$

(1.55)

It is easier to highlight important aspects of neutrino oscillations in a two flavour oscillation scenario. In such case, the mixing matrix parametrised by a mixing angle θ is given by

$$U = \begin{pmatrix} \cos\theta & \sin\theta \\ -\sin\theta & \cos\theta \end{pmatrix}$$

(1.56)

with

$$\begin{pmatrix} \nu_e \\ \nu_\mu \end{pmatrix} = U \cdot \begin{pmatrix} \nu_1 \\ \nu_2 \end{pmatrix}.$$

(1.57)

In this case, Eq. (1.54) assumes a simpler form:

$$P_{e\mu} = \sin^2 2\theta \sin^2 \left(1.27 \frac{\Delta m^2}{\rm eV^2} \frac{\rm GeV}{E} \frac{L}{\rm km} \right).$$

(1.58)

with $\Delta m^2 \equiv |m_2^2 - m_1^2|$. This demonstrates that a ν_e of energy E, might oscillate into a ν_μ after travelling a distance L, only if both θ and Δm^2 are non-zero. In the same spirit, one can rewrite Eq. (1.55) as

$$L_{\rm osc} = 2.48 \frac{E}{\rm GeV} \frac{\rm eV^2}{\Delta m^2} \; \rm km.$$

(1.59)

Table 1.1 Three neutrino mixing parameters [40]

$\sin^2 \theta_{12}$	0.307 ± 0.013
Δm_{21}^2	$(7.53 \pm 0.18) \times 10^{-5}\,\text{eV}^2$
$\sin^2 \theta_{23}$	0.545 ± 0.021
Δm_{32}^2	$0.002453 \pm 0.000034\,\text{eV}^2$
$\sin^2 \theta_{13}$	0.0218 ± 0.0007
δ_{13}	$1.36 \pm 0.17\,\pi\,\text{rad}$

In the case of atmospheric neutrinos, which are predominantly ν_μ, oscillating into ν_τ, one can get a rough estimate using this equation. Atmospheric neutrino spectrum peaks around a GeV. Taking $E \sim 1\,\text{GeV}$ and $\Delta m_{32}^2 \sim 10^{-3}\,\text{eV}^2$, $L_{\text{osc}} \sim 10^3\,\text{km}$. This means that studying neutrinos of energy around a GeV or so, which are produced at an atmospheric height of $\sim 10^3\,\text{km}$, Δm_{32}^2 can be probed at a level of $\sim 10^{-3}\,\text{eV}^2$. One can appreciate this impressive sensitivity keeping in mind that neutrino oscillation arises out of the interference of neutrino wave functions, like in the Michelson-Morley experiments with electromagnetic waves or in its modern incarnation, the gravitational wave detectors.

In a three neutrino oscillation scenario, the PMNS mixing matrix is a bit more involved. It is customary to parametrise such a 3×3 unitary matrix with three mixing angles θ_{12}, θ_{23} and θ_{13} and a Dirac CP violating phase δ_{13}, as in the CKM mixing matrix in the quark sector. A standard convention used for such parametrisation is

$$
\begin{aligned}
U &= \begin{pmatrix} U_{e1} & U_{e2} & U_{e3} \\ U_{\mu 1} & U_{\mu 2} & U_{\mu 3} \\ U_{\tau 1} & U_{\tau 2} & U_{\tau 3} \end{pmatrix} \\
&= \begin{pmatrix} 1 & 0 & 0 \\ 0 & c_{23} & s_{23} \\ 0 & -s_{23} & c_{23} \end{pmatrix} \begin{pmatrix} c_{13} & 0 & s_{13}e^{i\delta_{13}} \\ 0 & 1 & 0 \\ -s_{13}e^{-i\delta_{13}} & 0 & c_{13} \end{pmatrix} \begin{pmatrix} c_{12} & s_{12} & 0 \\ -s_{12} & c_{12} & 0 \\ 0 & 0 & 1 \end{pmatrix} \\
&= \begin{pmatrix} c_{12}c_{13} & s_{12}c_{13} & s_{13}e^{-i\delta_{13}} \\ -s_{12}c_{23} - c_{12}s_{23}s_{13}e^{i\delta_{13}} & c_{12}c_{23} - s_{12}s_{23}s_{13}e^{i\delta_{13}} & s_{23}c_{13} \\ s_{12}s_{23} - c_{12}c_{23}s_{13}e^{i\delta_{13}} & -c_{12}s_{23} - s_{12}c_{23}s_{13}e^{i\delta_{13}} & c_{23}c_{13} \end{pmatrix} \quad (1.60)
\end{aligned}
$$

where $c_{ij} \equiv \cos \theta_{ij}$ and $s_{ij} \equiv \sin \theta_{ij}$.

In the neutrino oscillation experiments, with solar, atmospheric, reactor or neutrino beam neutrinos, one can combine neutrino oscillation probabilities at different L and E, to determine the neutrino oscillation parameters Δm_{ij}^2, θ_{ij} and δ_{13}. Neutrino oscillation implies the existence of non-zero neutrino mass, which points at physics beyond the standard model of particle physics. Thus the parameters in the neutrino mixing matrix need to be determined from experiments only. The present estimates of these parameters are given in Table 1.1. The fit differs depending on the assumed hierarchy of the neutrino masses. See Ref. [40] for a review.

Probability oscillation formula for anti-neutrinos can be obtained from the same for neutrinos using CP transformation, which amounts to complex conjugation of the PMNS matrix elements used in Eq. (1.53). For $\bar{\nu}_\alpha \to \bar{\nu}_\beta$ oscillation, the probability is given by

$$P_{\bar{\alpha}\bar{\beta}} = \sum_{i,j} U_{\alpha i} U_{\beta i}^\star U_{\alpha j}^\star U_{\beta j} \exp\left[-i\frac{\Delta m_{ij}^2}{2E} L \right]. \tag{1.61}$$

Note that the requirement of CPT symmetry demands that oscillation probabilities of $\nu_\alpha \to \nu_\beta$ and $\bar{\nu}_\beta \to \bar{\nu}_\alpha$ are the equal. High energy astrophysical neutrinos can be useful in checking if such a symmetry is conserved in Nature.

Extending the discussion that follows Eq. (1.59), one can see that for a muon neutrino of energy 100 TeV, $L_{\text{osc}} \sim 10^8$ km, which is quite small compared to the distance of the prospective candidates for the sources of such neutrinos. For example, the nearest quasar from us, Markarian 231, is located 581 million light-years away. Thus, for astrophysical neutrinos $L_{\text{osc}} \ll L$, where L is the distance travelled. In this limit, the neutrino oscillates many times before its detection, so that the oscillatory terms in Eq. (1.54) get averaged out to zero. It leads to a simpler form for oscillation probability, applicable for both neutrinos and anti-neutrinos:

$$P_{\alpha\beta} = \sum_i |U_{\alpha i}|^2 |U_{\beta i}|^2. \tag{1.62}$$

1.6.2 *Neutrino Oscillations in Matter*

Usually the neutrinos are expected to get produced in astrophysical objects where the matter density is rather low. Hence, the vacuum neutrino oscillation formula Eq. (1.62) works quite well in most scenarios. However, the neutrinos might also pass through dense objects and undergo charged current coherent scattering with the electrons in them. In non-standard scenarios neutrinos may even interact with dark matter. Such interactions lead to modifications in the oscillation probabilities.

In a standard scenario, denoting the electron number density in the material as n_e and G_F as Fermi constant, θ and Δm^2 for two neutrino oscillation get modified in a medium as

$$\tan 2\theta_M = \frac{\Delta m^2 \sin 2\theta}{\Delta m^2 \cos 2\theta - A}, \tag{1.63}$$

and

$$\Delta m_M^2 = \sqrt{(\Delta m^2 \cos 2\theta - A)^2 + (\Delta m^2 \sin 2\theta)^2} \tag{1.64}$$

with $A = 2\sqrt{2} G_F n_e E$. As A depends both on E and n_e, a neutrino on its way through a medium may encounter an electron density for which $A = \Delta m^2 \cos 2\theta$, making the mixing angle in matter maximal, $\theta_M = \pi/4$ for any value of the mixing angle θ in vacuum. This is known as MSW resonance [41–43], named after Mikheev, Smirnov, and Wolfenstein. One can plug in Eqs. (1.63) and (1.64) in Eq. (1.58) to calculate the modified probabilities. For anti-neutrinos the sign of A is reversed.

One can generalise a two-neutrino oscillation in matter scenario to a three neutrino case. For high energy astrophysical neutrinos Eq. (1.62) will be modified accordingly. It is also possible that the neutrino undergoes non-adiabatic flavour transitions. However, here we will not elaborate on such issues as for the neutrinos of our interest, they are assumed to get produced in astrophysical objects where the matter density is usually taken to be negligible. Then the neutrinos pass essentially through a vacuum. Before reaching the detector, it passes through the Earth, and the length traversed through such a medium depends on the angle of incidence. However, one can convince oneself from Eq. (1.59) that the diameter of the Earth is too small compared to the oscillation length so that high energy neutrino oscillations within Earth can be neglected. In the case of non-standard interactions of the neutrinos, however, the situation could be significantly different.

1.6.3 Neutrino Flavour Ratios at Earth

As mentioned earlier, it is likely that the neutrinos are produced at the source with a flavour ratio $\nu_e : \nu_\mu : \nu_\tau \equiv 1{:}2{:}0$. But deviations from such a ratio is not impossible. Hence, here we will work with a generic ratio $f_e^S : f_\mu^S : f_\tau^S$ at the source. Then the ensuing flavour component at the detector is given by

$$f_\alpha^D - \sum_\beta P_{\alpha\beta} f_\beta^S, \tag{1.65}$$

where $\alpha, \beta = e, \mu, \tau$.
For the initial flavour ratio $f_e^S : f_\mu^S : f_\tau^S = 1{:}2{:}0$,

$$f_\alpha^D = \sum_i |U_{\alpha i}|^2 \left(|U_{ei}|^2 + 2|U_{\mu i}|^2 \right) \tag{1.66}$$

Plugging in the values of the parameters as in Table 1.1 in Eq. (1.60) one can verify that $|U_{\mu i}|^2 \simeq |U_{\tau i}|^2$. Together with the unitarity property of the U matrix, from Eq. (1.66) it follows that $f_e^D : f_\mu^D : f_\tau^D \simeq 1{:}1{:}1$. Exact equality is possible only for vanishing θ_{13}, which is not the case. Hence, in the standard scenario, one expects that if the neutrinos are produced in the astrophysical objects at a ratio 1:2:0, at earth all flavours will reach in almost equal proportions, irrespective of the energy. The same is also true for anti-neutrinos.

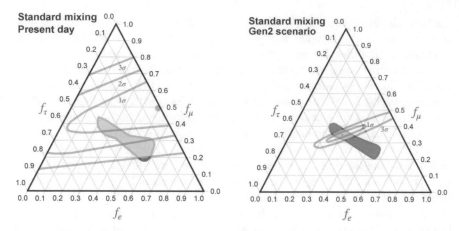

Fig. 1.5 Neutrino flavour composition at earth in the standard scenario. Left pane is drawn using the measured neutrino mixing parameters and IceCube constraints at present. The right pane uses projected sensitivities for Gen2. In both panes the light-green regions denote the allowed region given the uncertainties of the neutrino mixing parameters. For Gen2 a tentative set of such parameters is assumed. The best fit points are marked with dots. Darkened light-green regions indicate areas of exclusion at 3σ. Reproduced with permission from Ref. [44]

To probe the flavour ratio experimentally, one needs to take into account the uncertainties in the measured neutrino mixing parameters. Also we are not sure about the flavour ratio at the source. This also poses another source of uncertainty, which can be parametrised as $f_e^S : f_\mu^S : f_\tau^S = x : 1-x : 0$ where x lies between 0 and 1. In Fig. 1.5 the expected flavour ratios incorporating all such uncertainties are presented inside a flavour triangle as light-green shaded region for IceCube (left pane) and IceCube-Gen2 (right pane). The region for Gen2 demands use of projected uncertainties of neutrino parameters and in the right pane, the following parameters are used [44]: $s_{12}^2 = 0.306 \pm 0.002, s_{23}^2 = 0.441 \pm 0.01, s_{13}^2 = 0.0217 \pm 0.0005$ and $\delta_{13} = 261° \pm 15°$. In the left pane, the confidence level contours and the best fit point from IceCube measurements are shown following Ref. [45]. In the right pane expected sensitivities [46] of Gen2 for 15 years of observation has been used. It is evident that Gen2 would be instrumental in severely restricting the parameter space.

1.6.4 UHE Neutrino Propagation Through the Earth

As neutrinos pass through the earth, they undergo charged- and neutral-current interactions with earth's matter. In a charged-current interaction, a charged lepton is produced corresponding to the flavour of the neutrino. The electrons and muons get absorbed in the medium. The tauons, being more massive, can decay before they can get absorbed by the medium. Its decay $\tau \to \nu_\tau + X$, goes predominantly (65%) in the hadronic channel producing pions and the rest in the leptonic channel. The produced ν_τ can then undergo CC interaction to continue the cycle as $\nu_\tau \rightarrow \tau \rightarrow \nu_\tau \rightarrow \tau \rightarrow \cdots$. Thus the ν_τ can ultimately reach the detector, which may not be possible for ν_e and ν_μ. The same holds good for anti-neutrinos. So the tauon neutrino propagation needs to be dealt separately than the neutrinos of other flavours.

Neutrinos of all flavour undergo NC interactions producing another neutrino of same flavour but with degraded energy. So during propagation, due to NC interactions, neutrinos are not lost, but they shift to a lower energy bin in the spectrum. This effect is known as regeneration.

1.6.4.1 ν_e and ν_μ Propagation

For ν_e and ν_μ, propagation through the earth is dictated by the following differential equation,

$$\frac{\partial F_\nu(E, X)}{\partial X} = -\frac{F_\nu(E, X)}{\lambda_\nu(E)} + \frac{1}{\lambda_\nu(E)} \int_0^1 \frac{dy}{1 - y} K_\nu^{NC}(E, y) F_\nu(E_y, X). \quad (1.67)$$

where $F_\nu \equiv d\phi_\nu/dE$ is the differential neutrino flux.

The first term on the right hand side represents attenuation of neutrino flux due to absorption as it propagates through the earth. $\lambda_\nu = 1/(N_A \sigma_{\nu N}^{tot})$ denotes the interaction length for the neutrino with N_A as the Avogadro's number and $\sigma_{\nu N}^{tot} = \sigma_{\nu N}^{CC} + \sigma_{\nu N}^{NC}$ as the total neutrino nucleon cross-section (see Sect. 1.7).

The last term on the right hand side of Eq. (1.67) takes care of neutrino regeneration. E is the initial neutrino energy and y denotes the fraction of energy loss as the neutrino gets regenerated and $E_y = E/(1 - y)$. In the above,

$$K_\nu^{NC}(E, y) = \frac{1}{\sigma_{\nu N}^{tot}(E)} \frac{d\sigma_{\nu N}^{NC}(E_y, y)}{dy}. \quad (1.68)$$

As the neutrino traverses a distance L inside the earth to reach the detector, it crosses different material of varying density. So it is customary to define an effective length, as

$$X(\theta) = \int_0^L \rho(L') dL' \quad (1.69)$$

with $L = 2R_\oplus \cos\theta$, where R is the radius of the earth. θ is the nadir angle, where a vanishing θ implies that the neutrino is travelling through the diameter. X can be expressed in units of $g\,cm^{-2} = cm$ we. 'we' stands for 'water equivalent' that indicates the equivalent length traversed in water.

Table 1.2 Earth density profile according to the Preliminary Reference Earth Model [47]

r (km)	$\rho(x = r/R_\oplus)\,(\mathrm{g\,cm^{-3}})$	
$r < 1221.5$	$13.0885 - 8.8381x^2$	Core
$1221.5 < r < 3480$	$12.5815 - 1.2638x - 3.6426x^2 - 5.5281x^3$	
$3480 < r < 5701$	$7.9565 - 6.4761x + 5.5283x^2 - 3.0807x^3$	Mantle
$5701 < r < 5771$	$5.3197 - 1.4836x$	
$5771 < r < 5971$	$11.2494 - 8.0298x$	
$5971 < r < 6151$	$7.1089 - 3.8045x$	
$6151 < r < 6346.6$	$2.691 + 0.6924x$	
$6346.6 < r < 6356$	2.900	Crust
$6356 < r < 6368$	2.600	
$6368 < r < R_\oplus = 6371$	1.020	Ocean

The best knowledge about ρ, the density of earth at different distances from the centre, comes from the Preliminary Reference Earth Model (PREM) [47]. In this model, earth is approximated as a spherically symmetric ball that has varying density along its radius as presented in Table 1.2.

In absence of the regeneration term, Eq. (1.67) can have a simple analytical solution [48]

$$F_\nu(E, X) = F_\nu^0(E) \exp\left[-\frac{X}{\lambda_\nu(E)}\right] \qquad (1.70)$$

where, $F_\nu^0(E) \equiv F_\nu(E, X = 0)$ denotes the initial neutrino flux. The regeneration term modifies the interaction length $\lambda_\nu(E)$ to an effective interaction length $\Lambda_\nu(E, X)$. Such a solution can be obtained via a semi-analytical method [49]. Alternatively, one can solve it numerically. See Refs. [50, 51] for a recent exposure.

1.6.5 ν_τ Propagation

As mentioned earlier, due to the $\nu_\tau \to \tau \to \nu_\tau \to \tau \to \cdots$ chain, the propagation equations get more involved. One needs to write the following coupled differential equations [52, 53] that can describe propagation of both ν_τ and τ in tandem.

$$\frac{\partial F_{\nu_\tau}(E, X)}{\partial X} = -\frac{F_{\nu_\tau}(E, X)}{\lambda_\nu(E)} + \frac{1}{\lambda_\nu(E)} \int_0^1 \frac{dy}{1-y} K_\nu^{NC}(E, y) F_{\nu_\tau}(E_y, X)$$

$$+ \int_0^1 \frac{dy}{1-y} K_\tau(E, y) F_\tau(E_y, X), \qquad (1.71)$$

$$\frac{\partial F_\tau(E, X)}{\partial X} = -\frac{F_\tau(E, X)}{\hat{\lambda}(E)} + \frac{\partial \left[\gamma(E) F_\tau(E, X) \right]}{\partial E}$$

$$+ \frac{1}{\lambda_\nu(E)} \int_0^1 \frac{dy}{1-y} K_\nu^{CC}(E, y) F_{\nu_\tau}(E_y, X). \qquad (1.72)$$

with,

$$K_\tau(E, y) = \frac{1}{\lambda_\tau(E)} K_\tau^{CC}(E, y) + \frac{1}{\lambda_\tau^{dec}(E)} K_\tau^{dec}(E, y) \qquad (1.73)$$

where,

$$K_\tau^{CC}(E, y) = \frac{1}{\sigma_{\tau N}^{tot}(E)} \frac{d\sigma_{\tau N}^{CC}(E_y, y)}{dy}, \quad K_\tau^{dec}(E, y) = \frac{1}{\Gamma_\tau^{tot}(E)} \frac{d\Gamma_{\tau \to \nu_\tau X}(E_y, y)}{dy}$$

with,

$$\frac{1}{\lambda_\tau} = N_A \sigma_{\tau N}^{tot}, \quad \frac{1}{\hat{\lambda}} = \frac{1}{\lambda_\tau^{CC}} + \frac{1}{\lambda_\tau^{dec}} \quad \text{and} \quad \frac{1}{\lambda_\tau^{CC}} = N_A \sigma_{\tau N}^{CC}. \qquad (1.74)$$

The decay length of the $\iota^\pm$ is

$$\lambda_\tau^{dec}(E, X, \theta) = \frac{E}{m_\tau} c\tau_\tau \rho(X, \theta), \qquad (1.75)$$

with $m_\tau = 1.777\,\text{GeV}$, $c\tau_\tau = 87.11\,\mu\text{m}$. To a good approximation, $\rho(X, \theta)$ can be replaced with an averaged density of earth along the direction of propagation of the neutrino defined as [53]

$$\rho_{avg}(\theta) = \frac{1}{L} \int_0^L \rho(r(z, \theta))\,dz = \frac{X(\theta)}{L}, \qquad (1.76)$$

with $r(z, \theta) = \sqrt{R_\oplus^2 + z^2 - zL}$ and $L = 2R_\oplus \cos\theta$.

Here, both F_{ν_τ} and F_τ denote differential fluxes as before. In Eq. (1.71), apart from the attenuation and regeneration terms, there is also a term accounting for ν_τ generation from τ decay.

In Eq. (1.72), the first term on the right hand side is the usual attenuation term from τ absorption in the medium. The next term takes care of tau energy loss with $\gamma(E) = -dE/dX = \alpha + \beta E$. For values of α and β, see Ref. [53] and the references therein. For $E < 10^8$ GeV, the energy loss of taus can be neglected. Above this energy the energy loss restricts the growth of λ_τ^{dec} with energy. λ_τ^{CC} decreases with energy due to the enhancement of cross-section. Around $E \sim 10^{12}$ GeV, λ_τ^{dec} saturates and becomes comparable in value with λ_τ^{CC}. The last term accounts for τ production from ν_τ via CC interaction.

Equations (1.71) and (1.72) can be solved semi-analytically or numerically to find the resulting ν_τ flux arriving at the detector after passing through the earth, given some initial flux.

1.7 Neutrino Cross-Sections

Astrophysical neutrinos can undergo weak interaction with the protons, neutrons and electrons in the matter as they pass through. Compared to the low energy neutrinos, like solar or atmospheric neutrinos, the neutrino-nucleon (νN) cross-section for high energy astrophysical neutrinos is significantly on a higher side. As a result, for extremely energetic ones, the Earth is no longer transparent. For neutrino energies of more than a few PeVs, νN cross-sections have significant theoretical and experimental uncertainties. Detection of such neutrinos can thus improve our understanding of νN cross-sections at extreme energies, that will in turn help us probe nature of QCD at these energies and other non-standard theoretical propositions. We will also comment on their interactions with the electrons.

High energy astrophysical neutrinos can interact with a stationary nucleon and undergo deep inelastic scattering (DIS) as at these energies it can see its constituents – the partons. Such interactions can take place in a t-channel charged-current (CC) process through an exchange of a W boson, $\nu_l N \rightarrow l X$, producing a lepton of corresponding flavour and some hadronic remnants. If it occurs via the exchange of a Z boson, $\nu_l N \rightarrow \nu_l X$, then the process is known as a neutral-current (NC) process, and a neutrino of same flavour comes out. See Fig. 1.6 for the corresponding Feynman diagrams.

To define the kinematic variables[1] pertaining to a deep inelastic scattering process let us consider the following one:

$$\nu_l(k) + N(P) \rightarrow l(k') + X(p_X). \tag{1.77}$$

From Fig. 1.7, it can be seen that the neutrino with momentum k interacts with a parton of momentum p that carries an x fraction of the momentum P of the parent nucleon, $p = xP$. In the parton model, quarks are treated as a free Dirac particle, which allows us to neglect the mass of the quarks. Therefore, $(p + q)^2 = 0$ and

---

[1] See [54, 55] for further reading.

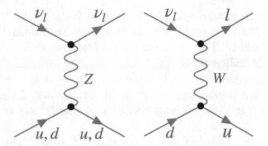

Fig. 1.6 Leading order Feynman diagrams for neutrino quark interaction. Similar diagrams can be drawn for anti-neutrinos by changing particles in these diagrams into their anti-particles. Here u and d stand for up- and down-type quarks for all generations

Fig. 1.7 Charged current νN deep inelastic scattering. Incoming nucleon can be a proton or a neutron. For a neutral current scattering, the outgoing neutrino would be a neutrino of the same flavour. All such parton level processes are shown in Fig. 1.6

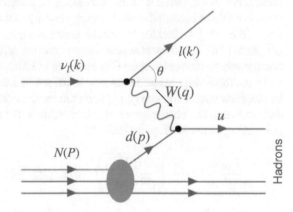

$p^2 = 0$. These help us to express the dimensionless Lorentz scalar x, known as Bjorken-x as

$$x = \frac{-q^2}{2P.q}. \tag{1.78}$$

Note that $0 \le x \le 1$.

The invariant mass W of the hadronic system is given by $W^2 = (P + q)^2$. Another Lorentz-invariant quantity, the squared centre-of-mass energy can be written as

$$s = (k + P)^2 = M^2 + 2k \cdot P, \tag{1.79}$$

where M represents the mass of the nucleon.

The inelasticity parameter y, defined as

$$y = \frac{q \cdot P}{k \cdot P}, \tag{1.80}$$

takes a simpler form $y = (E - E')/E$ in the rest frame of the nucleon. y lies in between 0 and 1. Here, the energies of the incoming neutrino and outgoing lepton are denoted by E and E'. Thus, y stands for the fractional energy loss of the lepton due to this inelastic collision.

The Mandlestam variable t can be written as $t = q^2 = (k - k')^2 \simeq -2k \cdot k'$. It is easy to check the overall sign of q^2 in the rest frame of the nucleon: $q^2 = -4EE' \sin^2(\theta/2) \leq 0$. Hence, it is customary to define $Q^2 \equiv -q^2$, that makes Q^2 positive.

For deep inelastic scattering, $Q^2 \gg M^2$ and $W^2 \gg M^2$. The limit $Q^2 \gg M^2$ implies that the scattering is 'deep', that means the neutrino can resolve the partons inside the nucleon. $W^2 \gg M^2$ reflects extreme inelasticity. The four Lorentz invariant quantities s, Q^2, x and y are not independent. They are related as $Q^2 \simeq xys$ for DIS. Hence, besides s, two out of the remaining variables can be chosen as independent ones. For DIS, s is related to the incoming energy of the neutrino by a simple formula $s = 2ME$. To get a feel for the numbers, one can note that the centre-of-mass energy $\sqrt{s}$ for a 1 PeV astrophysical neutrino interacting with a nucleon at rest is ~ 1 TeV, comparable to the parton level CM energies at LHC.

In the following, we will consider neutrinos interacting with material dominated by isoscalar targets like oxygen, silicon nucleus, containing equal number of protons and neutrons. The charged current contribution to the neutrino nucleon scattering, at the leading order, is given by [56–58]

$$\frac{d^2\sigma_{CC}}{dQ^2 dx} = \frac{G_F^2}{\pi} \left(\frac{m_W^2}{Q^2 + m_W^2} \right)^2 \left(q(x, Q^2) + \bar{q}(x, Q^2)(1 - y^2) \right), \qquad (1.81)$$

where $q(x, Q^2) = f_d + f_s + f_b$ and $\bar{q}(x, Q^2) = f_{\bar{u}} + f_{\bar{c}} + f_{\bar{t}}$. Here, f_q and $f_{\bar{q}}$ with q running over all flavours stand for parton distribution functions (PDF). PDFs are functions of x and Q^2 and are calculated by fitting data from fixed target experiments and collider experiments such as HERA (ep scattering), Tevatron ($\bar{p}p$ scattering) and LHC (pp scattering). G_F is the Fermi coupling constant. Here, the effect of isoscalar targets can be incorporated by averaging over isospin, that amounts to the following replacements: $f_{u/d} \rightarrow (f_u + f_d)/2$ and $f_{\bar{u}/\bar{d}} \rightarrow (f_{\bar{u}} + f_{\bar{d}})/2$ cross-sections for anti-neutrinos can be obtained by the interchange $f_q \leftrightarrow f_{\bar{q}}$.

The leading order neutral current cross-section can be written in a similar manner.

$$\frac{d^2\sigma_{NC}}{dQ^2 dx} = \frac{G_F^2}{\pi} \left(\frac{m_Z^2}{Q^2 + m_Z^2} \right)^2 \left(q^0(x, Q^2) + \bar{q}^0(x, Q^2)(1 - y^2) \right). \qquad (1.82)$$

where,

$$q^0 = (f_u + f_c + f_t)L_u^2 + (f_{\bar{u}} + f_{\bar{c}} + f_{\bar{t}})R_u^2 + (f_d + f_s + f_b)L_d^2$$
$$+ (f_{\bar{d}} + f_{\bar{s}} + f_{\bar{b}})R_d^2,$$

and

$$\bar{q}^0 = (f_u + f_c + f_t)R_u^2 + (f_{\bar{u}} + f_{\bar{c}} + f_{\bar{t}})L_u^2 + (f_d + f_s + f_b)R_d^2$$
$$+ (f_{\bar{d}} + f_{\bar{s}} + f_{\bar{b}})L_d^2.$$

with

$$L_u = \frac{1}{2} - \frac{2}{3}\sin^2\theta_W, \quad L_d = -\frac{1}{2} + \frac{1}{3}\sin^2\theta_W,$$

$$R_u = -\frac{2}{3}\sin^2\theta_W, \quad R_d = \frac{1}{3}\sin^2\theta_W.$$

At the next-to-leading-order, these expressions should be convoluted with appropriate coefficient functions to get the structure functions [51, 58].

At high energies, the W or Z propagator restricts Q^2 from going beyond $M_{W,Z}^2$. As a result, from the relations $Q^2 = xys$ and $y \leq 1$ it follows that, x falls with neutrino energy as $x \gtrsim M_W^2/(2ME)$. At $E \sim$ EeV energies, this demands precise knowledge of the PDFs for x as small as $\sim 10^{-5}$, or even less. However, around $Q^2 \sim 10^4$ GeV^2, the current experiments probe $x \gtrsim 10^{-4}$. So to compute neutrino nucleon cross-sections for $E \gtrsim 100$ PeV, one needs to extrapolate PDFs to smaller x, a regime not explored by current experiments. This leads to an uncertainty in the cross-sections (see Fig. 1.8).

At low x and Q^2 values, the PDF uncertainties are significant. The dominant contributions do not come from very low Q^2 values. This ensures that the sensitivity of the neutrino cross-section on PDF uncertainties is rather small even at $E \sim 1$ ZeV [57]. However, as shown in Ref. [59], that uses MSTW 2008 PDF, the uncertainties in the cross-sections can be larger. Figure 1.8 illustrates uncertainties in cross-sections as presented by various groups. It has been argued in Ref. [59] that the difference in the uncertainties stems from the fact that in Ref. [57] the gluon distribution is parametrised as $g(x) \propto x^\delta$, whereas MRST 2008 PDF uses a different parametrisation: $xg(x) \propto A_1 x^{\delta_1} + A_2 x^{\delta_2}$.

A rough numerical estimate of total neutrino cross-sections, accurate to within $\sim 10\%$ in the energy range 10^7 GeV $\geq E \leq 10^{12}$ GeV, was updated in Refs. [57, 63] as

$$\log_{10}\left(\frac{\sigma_{CC}}{cm^2}\right) = -39.59 \left[\log_{10}\left(\frac{E_\nu}{GeV}\right)\right]^{-0.0964}, \tag{1.83}$$

$$\log_{10}\left(\frac{\sigma_{NC}}{cm^2}\right) = -40.13 \left[\log_{10}\left(\frac{E_\nu}{GeV}\right)\right]^{-0.0983}. \tag{1.84}$$

Neutrino-electron cross-sections are usually small compared to the neutrino-nucleon cross-sections, due to the small mass of the electron (Fig. 1.9). However, for processes like $\bar{\nu}_e e \to$ hadrons, $\bar{\nu}_e e \to \bar{\nu}_e e$, $\bar{\nu}_e e \to \bar{\nu}_\mu \mu$ and $\bar{\nu}_e e \to \bar{\nu}_\tau \tau$, that proceed in the s-channel interaction via exchange of a W boson, one can see that

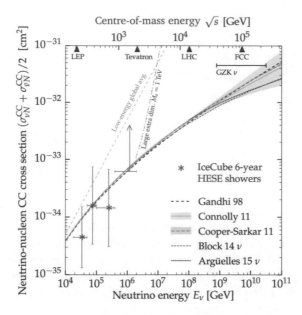

Fig. 1.8 νN charged current cross-section, averaged over neutrinos and anti-neutrinos. PDF uncertainties are shown as bands. Results from different groups are shown: Gandhi 98 [56], Connolly 11 [59], Cooper-Sarkar 11 [58], Block 14 ν [60], Arguelles 15 ν [61]. CC cross-section estimates [62] using HESE events in the 6-year data from IceCube are also indicated. Reproduced with permission from Ref. [62]

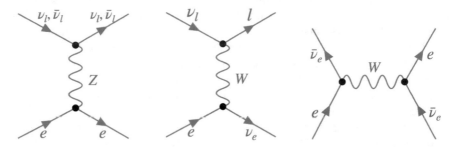

Fig. 1.9 Leading order Feynman diagrams for neutrino or anti-neutrino scattering with electrons. Neutrinos and anti-neutrinos of all flavours undergo neutral current interaction. Neutrinos of all flavours participate in the charged current interaction with the electron. But only $\bar{\nu}_e$ can interact with the electrons via an s-channel neutral current interaction due to the absence of μ and τ in the medium

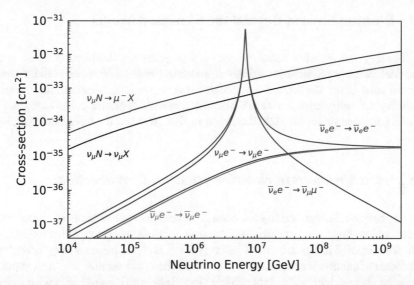

Fig. 1.10 Neutrino-nucleon and neutrino-lepton total scattering cross-sections plotted against incoming neutrino energy. Neutrino-nucleon cross-sections are consistent with Ref. [58] that used next-to-leading-order (NLO) HERA1.5 parton distribution functions and massless coefficient functions at NLO. The Glashow resonance in $\bar{\nu}_e e^-$ channels are clearly evident around $E = 6.3\,$PeV. Figure courtesy: Swapnesh Khade

at $\sqrt{s} = \sqrt{2Em_e} = M_W$ a resonance should occur, increasing the cross-section more than two orders of magnitude. It translates to the incoming neutrino energy $E = 6.3\,$PeV (see Fig. 1.10). The resonance is named after Glashow [64]. Contributions to the cross-section is more pronounced for the energy of the neutrino in the range 5.7–7 PeV, as decided by the full-width-at-half-maxima of the resonance: from $(M_W - 2\Gamma_W)^2/(2m_e)$ to $(M_W + 2\Gamma_W)^2/(2m_e)$. IceCube collaboration has recently reported first observation of such a resonance in Ref. [65].

The cross-section for various neutrino-electron scattering processes are given by Ref. [66]

$$\frac{d\sigma(\bar{\nu}_e e \to \text{hadrons})}{dy} = \frac{d\sigma(\bar{\nu}_e e \to \bar{\nu}_\mu \mu)}{dy} \cdot \frac{\Gamma(W \to \text{hadrons})}{\Gamma(W \to \mu\bar{\nu}_\mu)} \qquad (1.85)$$

where,

$$\frac{d\sigma(\bar{\nu}_e e \to \bar{\nu}_\mu \mu)}{dy} = \frac{G_F^2 m_e E}{2\pi} \frac{4(1-y)^2[1 - (m_\mu^2 - m_e^2)/(2m_e E)]^2}{(1 - 2m_e E/M_W^2)^2 + \Gamma_W^2/M_W^2}. \qquad (1.86)$$

Here, $y = E'/E$, the ratio of the energies of the outgoing lepton and incoming neutrino. $\Gamma_W = 2.1\,$GeV is the total decay width of W. $\Gamma(W \to \text{hadrons}) = 1.4\,$GeV and $\Gamma(W \to \mu\bar{\nu}_\mu) = 0.2\,$GeV denote the partial decay widths in these channels.

1.8   Probing Physics Beyond the Standard Model

High energy astrophysical neutrinos can help us probe physics beyond the standard model of particle physics, sometimes surpassing the limitations of collider based experiments. Given the limited scope of this review, we will discuss some scenarios focussing on high energy (>100 TeV) neutrinos, possibly of extra-galactic origin. For a more complete review see the discussion in Ref. [67] and the references therein.

1.8.1   Modifications in Neutrino-Nucleon Cross-Sections

Physics beyond SM can modify the neutrino-nucleon cross-sections at higher energies. For ultra-high energy neutrinos such cross-sections receive significant contributions from small Bjorken $x \lesssim 10^{-4}$. For $x \lesssim 10^{-2}$, with the increase in Q^2, the gluon distribution function increases, and as a result, the cross-section increases rapidly. However, such an increase cannot continue indefinitely as it would otherwise violate unitarity [68]. For small x, gluons should get very tightly packed, thereby showing a shadowing effect. Moreover, as the neutrino energy increases, these gluons are pushed to attain larger momenta, resulting in weaker coupling. Thus at some point, the gluon density should saturate forming a colour glass condensate, thereby reducing the cross-section at higher energies [69]. Such a behaviour is yet to be observed in experiments. Collisions of UHE neutrinos with nucleons in the detectors can help us probe such a trend in the νN cross-sections.

Microscopic black holes (MBH) can get produced in particle collisions, if the impact factor is less than the event horizon of the black holes [70]. In νN scattering, such TeV scale MBH can get produced, which gets evaporated with a rest lifetime of TeV^{-1}. If one neglects particle masses, then MBH decays democratically [71] into all SM degrees of freedom, with hadrons and muons carrying 75% and 3% of the MBH energy respectively. Such decays can thus not only lead to spectacular signatures in a detector like IceCube [72], but also enhance νN cross-sections [73].

The presence of large extra spatial dimensions can also lead to additional νN interactions via graviton exchanges, enhancing the cross-sections [48, 74, 75]. IceCube has already been able to put stringent constraints on such scenarios (see Fig. 1.9). Similarly, leptoquark mediated diagrams can also increase νN cross-sections [76].

Non-standard interactions of neutrinos with quarks are another source of νN cross-section enhancements. Observations at IceCube can compete with the experiments at LHC to constrain such possibilities. If such an interaction is mediated by the exchange of an additional neutral gauge boson, then in certain regions of the parameter space IceCube can be more effective than LHC [77].

In the SM, around the centre-of-mass energy ~ 9 TeV, non-perturbative effects mediated by sphaleron transitions are expected to show up. These processes violate baryon (B) and lepton (L) numbers, but $B - L$ remains conserved [78, 79]. Sphaleron induced neutrino-quark interactions can arise from gauge invariant operators like

$(\bar{q}\bar{q}\bar{q})_1(\bar{q}\bar{q}\bar{q})_2(\bar{q}\bar{q}\bar{q})_3(\bar{\ell}_1\bar{\ell}_2\bar{\ell}_3)$, leading to $qv \rightarrow 8\bar{q}2\bar{\ell}$ and other such processes with high multiplicity. So it may lead to spectacular signatures at the detectors and enhance vN cross-sections [80–82] for neutrinos of energy several PeVs or more.

Neutrinos can produce double muon events via trident [83] processes. New physics can also contribute to such processes. An active, light neutrino can interact with the detector to produce a sterile, heavy neutrino [84] that can also decay to produce a spectacular double muon event, as the vertex with these two muons would be spatially displaced from the vN vertex.

The present error in the measurement of vN cross-sections is about 30–40%, up to a PeV, considering one year data at IceCube [85]. In IceCube-Gen2, the statistics per year would improve approximately by one order of magnitude, and one might be able to measure cross-section beyond 10 PeV. With the present sensitivity of the IceCube, no event at such extreme energies could be detected. However, with Gen2 one can expect to detect about three events per year from cosmogenic neutrinos only, above 100 PeV [67]. Thus the new physics scenarios that lead to changes in the vN cross-sections at neutrino energies more than a PeV can be probed in experiments involving high energy astrophysical neutrinos.

1.8.2 Changes in Neutrino Flavour Ratios

In the standard scenario, consideration of initial flavour ratio as 1:2:0, and the neutrino mixing angles as extracted from low energy neutrino oscillation experiments lead to a flavour ratio at the earth 1:1:1, that is independent of neutrino energy. As has been indicated earlier, the initial flavour ratio may differ, and that would lead to a different flavour ratio at the earth. For example, in 'muon-damped' sources, the muons lose energy in the surrounding magnetic field before they decay, to produce neutrinos in the ratio 0:1:0. In 'neutron beam' sources, neutrinos are produced from beta decays, so the flavour ratio at source turns out to be 1:0:0. Similar to a neutrino factory on earth, 'muon beam' sources can also exist, where muons from higher energies can pile up at lower energies, producing a flavour ratio 1:1:0. Such non-standard scenarios can induce an energy dependence in flavour ratios [86, 87]. Neutrino flavour oscillations at ultra-high energies and over cosmological distances would anyway be better tested in IceCube-Gen2.

Even a feeble interaction of high energy neutrinos with the ultralight dark matter (DM) inside an astrophysical object might induce a matter effect that can lead to an energy dependence in the observed flavour ratio at earth. This can in turn be used to do neutrino astronomy, probing the solitonic structure formed by the Bose-Einstein condensed ultralight DM [88].

In general, new physics interactions that are not diagonal in flavour can alter the standard expectations in the neutrino flavour ratio at earth [89].

1.8.3   Testing Violations of Fundamental Symmetry

Often quantum gravity inspired scenarios accommodate a different space-time scenario near the Planck scale that allows for violations of fundamental symmetries like violation of Lorentz invariance (VLI), CPT, etc. VLI can lead to reduction of UHE neutrino flux and shift its peak [90]. High statistics of ultrahigh energy astrophysical neutrinos at the IceCube-Gen2 can help us constrain such BSM scenarios. The observation of the TXS 0506+056 event did already put stringent constraints on such BSM scenarios, once one assumes that the neutrino and gamma rays emitted simultaneously from the source [91, 92]. VLI can also induce a neutrino potential that ultimately affects the flavour composition at earth [93].

References

1. M. Tanabashi et al., Particle Data Group. Phys. Rev. D **98**(3), 030001 (2018)
2. M.S. Longair, *High Energy Astrophysics*, 3rd edn. (Cambridge University Press, 2011)
3. E. Fermi, Phys. Rev. **75**, 1169–1174 (1949)
4. W.I. Axford, E. Leer, G. Skadron, in *Proceedings of the 15th International Cosmic Ray Conference*, vol. 11, pp. 132–135 (1977)
5. G.F. Krymsky, Dokl. Akad. Nauk SSSR **234**, 1306–08 (1977)
6. R.D. Blandford, J.P. Ostriker, Astrophys. J. Lett. **221**, L29–L32 (1978)
7. A.R. Bell, Mon. Not. Roy. Astron. Soc. **182**, 147–156 (1978)
8. A.M. Hillas, Ann. Rev. Astron. Astrophys. **22**, 425–444 (1984)
9. K. Kotera, A.V. Olinto, Ann. Rev. Astron. Astrophys. **49**, 119–153 (2011). arXiv:1101.4256 [astro-ph.HE]
10. K.V. Ptitsyna, S.V. Troitsky, Phys. Usp. **53**, 691–701 (2010). arXiv:0808.0367 [astro-ph]
11. M. Ahlers, F. Halzen, Phys. Rev. D **90**(4), 043005 (2014). arXiv:1406.2160 [astro-ph.HE]
12. M. Ahlers, F. Halzen, PTEP **2017**(12), 12A105 (2017)
13. O. Pisanti, J. Phys. Conf. Ser. **1263**(1), 012004 (2019). arXiv:1906.12258 [astro-ph.CO]
14. L.A. Anchordoqui, T. Montaruli, Ann. Rev. Nucl. Part. Sci. **60**, 129–162 (2010). arXiv:0912.1035 [astro-ph.HE]
15. M. Ahlers, F. Halzen, Phys. Rev. D **86**, 083010 (2012). arXiv:1208.4181 [astro-ph.HE]
16. A. Aab et al. [Pierre Auger], JCAP **04** (2017), 038 [erratum: JCAP **03** (2018), E02]. arXiv:1612.07155 [astro-ph.HE]
17. E. Waxman, J.N. Bahcall, Phys. Rev. D **59**, 023002 (1999). arXiv:hep-ph/9807282 [hep-ph]
18. J.N. Bahcall, E. Waxman, Phys. Rev. D **64**, 023002 (2001). arXiv:hep-ph/9902383 [hep-ph]
19. M. Ahlers, F. Halzen, Prog. Part. Nucl. Phys. **102**, 73–88 (2018). arXiv:1805.11112 [astro-ph.HE]
20. I. Tamborra, S. Ando, K. Murase, JCAP **09**, 043 (2014). arXiv:1404.1189 [astro-ph.HE]
21. K. Greisen, Phys. Rev. Lett. **16**, 748–750 (1966)
22. G.T. Zatsepin, V.A. Kuzmin, JETP Lett. **4**, 78–80 (1966)
23. F.W. Stecker, Phys. Rev. **180**, 1264–1266 (1969)
24. D. Allard, M. Ave, N. Busca, M.A. Malkan, A.V. Olinto, E. Parizot, F.W. Stecker, T. Yamamoto, JCAP **09**, 005 (2006). arXiv:astro-ph/0605327 [astro-ph]
25. V. Berezinsky, A.Z. Gazizov, S.I. Grigorieva, Phys. Lett. B **612**, 147–153 (2005). arXiv:astro-ph/0502550 [astro-ph]
26. M. Ahlers, L.A. Anchordoqui, H. Goldberg, F. Halzen, A. Ringwald, T.J. Weiler, Phys. Rev. D **72**, 023001 (2005). arXiv:astro-ph/0503229 [astro-ph]

27. R. Aloisio, V. Berezinsky, P. Blasi, A. Gazizov, S. Grigorieva, B. Hnatyk, Astropart. Phys. **27**, 76–91 (2007). arXiv:astro-ph/0608219 [astro-ph]
28. M. Ahlers, L.A. Anchordoqui, M.C. Gonzalez-Garcia, F. Halzen, S. Sarkar, Astropart. Phys. **34**, 106–115 (2010). arXiv:1005.2620 [astro-ph.HE]
29. V.S. Berezinsky, G.T. Zatsepin, Phys. Lett. B **28**, 423–424 (1969)
30. M. Ahlers, Phys. Proc. **61**, 392–398 (2015)
31. A.A. Abdo et al., Fermi-LAT. Phys. Rev. Lett. **104**, 101101 (2010). arXiv:1002.3603 [astro-ph.HE]
32. K. Kotera, D. Allard, A.V. Olinto, JCAP **10**, 013 (2010). arXiv:1009.1382 [astro-ph.HE]
33. A.M. Hopkins, J.F. Beacom, Astrophys. J. **651**, 142–154 (2006). arXiv:astro-ph/0601463 [astro-ph]
34. R. Hill, K.W. Masui, D. Scott, Appl. Spectrosc. **72**(5), 663–688 (2018). arXiv:1802.03694 [astro-ph.CO]
35. A. Cooray, R. Soc, Open Sci. **3**, 150555 (2016). arXiv:1602.03512 [astro-ph.CO]
36. G. Breit, J.A. Wheeler, Phys. Rev. **46**(12), 1087–1091 (1934)
37. A. De Angelis, G. Galanti, M. Roncadelli, Mon. Not. Roy. Astron. Soc. **432**, 3245–3249 (2013). arXiv:1302.6460 [astro-ph.HE]
38. R.J. Gould, G.P. Schreder, Phys. Rev. **155**, 1408–1411 (1967)
39. A. De Angelis, M. Mallamaci, Eur. Phys. J. Plus **133**, 324 (2018). arXiv:1805.05642 [astro-ph.HE]
40. P.A. Zyla et al. [Particle Data Group], PTEP **2020**(8), 083C01 (2020)
41. L. Wolfenstein, Phys. Rev. D **17**, 2369–2374 (1978)
42. L. Wolfenstein, Phys. Rev. D **20**, 2634–2635 (1979)
43. S.P. Mikheyev, A.Y. Smirnov, Sov. J. Nucl. Phys. **42**, 913–917 (1985)
44. R.W. Rasmussen, L. Lechner, M. Ackermann, M. Kowalski, W. Winter, Phys. Rev. D **96**(8), 083018 (2017). arXiv:1707.07684 [hep-ph]
45. M.G. Aartsen et al., IceCube. Astrophys. J. **809**(1), 98 (2015). arXiv:1507.03991 [astro-ph.HE]
46. M. Kowalski, *The Icecube Particle Astrophysics Symposium (IPA)* (Madison, USA, 2017)
47. A.M. Dziewonski, D.L. Anderson, Phys. Earth Planet. Inter. **25**, 297–356 (1981)
48. K. Giesel, J.H. Jureit, E. Reya, Astropart. Phys. **20**, 335–360 (2003). arXiv:astro-ph/0303252 [astro-ph]
49. V.A. Naumov, L. Perrone, Astropart. Phys. **10**, 239–252 (1999). arXiv:hep-ph/9804301 [hep-ph]
50. A.C. Vincent, C.A. Argüelles, A. Kheirandish, JCAP **11**, 012 (2017). arXiv:1706.09895 [hep-ph]
51. A. Garcia, R. Gauld, A. Heijboer, J. Rojo, JCAP **09**, 025 (2020). arXiv:2004.04756 [hep-ph]
52. S.I. Dutta, M.H. Reno, I. Sarcevic, Phys. Rev. D **62**, 123001 (2000). arXiv:hep-ph/0005310 [hep-ph]
53. S. Rakshit, E. Reya, Phys. Rev. D **74**, 103006 (2006). arXiv:hep-ph/0608054 [hep-ph]
54. F. Halzen, A.D. Martin, *Quarks and Leptons: An Introductory Course in Modern Particle Physics*, 1st edn. (Wiley, 1984)
55. R. Devenish, A. Cooper-Sarkar, *Deep Inelastic Scattering*, 1st edn. (Oxford University Press, 2004)
56. R. Gandhi, C. Quigg, M.H. Reno, I. Sarcevic, Phys. Rev. D **58**, 093009 (1998). arXiv:hep-ph/9807264 [hep-ph]
57. A. Cooper-Sarkar, S. Sarkar, JHEP **01**, 075 (2008). arXiv:0710.5303 [hep-ph]
58. A. Cooper-Sarkar, P. Mertsch, S. Sarkar, JHEP **08**, 042 (2011). arXiv:1106.3723 [hep-ph]
59. A. Connolly, R.S. Thorne, D. Waters, Phys. Rev. D **83**, 113009 (2011). arXiv:1102.0691 [hep-ph]
60. M.M. Block, L. Durand, P. Ha, Phys. Rev. D **89**(9), 094027 (2014). arXiv:1404.4530 [hep-ph]
61. C.A. Argüelles, F. Halzen, L. Wille, M. Kroll, M.H. Reno, Phys. Rev. D **92**(7), 074040 (2015). arXiv:1504.06639 [hep-ph]
62. M. Bustamante and A. Connolly, Phys. Rev. Lett. **122**(4), 041101 (2019). arXiv:1711.11043 [astro-ph.HE]

63. M. Ahlers, K. Helbing, C. Pérez de los Heros, Eur. Phys. J. C **78**(11), 924 (2018). arXiv:1806.05696 [astro-ph.HE]
64. S.L. Glashow, Phys. Rev. **118**, 316–317 (1960)
65. M.G. Aartsen et al. [IceCube], Nature **591**(7849), 220–224 (2021) [erratum: Nature **592**(7855), E11 (2021)]
66. R. Gandhi, C. Quigg, M.H. Reno, I. Sarcevic, Astropart. Phys. **5**, 81–110 (1996). arXiv:hep-ph/9512364 [hep-ph]
67. M.G. Aartsen et al., IceCube-Gen2. J. Phys. G **48**(6), 060501 (2021). arXiv:2008.04323 [astro-ph.HE]
68. M. Froissart, Phys. Rev. **123**, 1053 (1961)
69. E.M. Henley, J. Jalilian-Marian, Phys. Rev. D **73**, 094004 (2006). arXiv:hep-ph/0512220 [hep-ph]
70. K.S. Thorne, Nonspherical gravitational collapse: a short review, in *Magic Without Magic— John Archibald Wheeler. A Collection of Essays in Honor of his 60th Birthday*. ed. by J.R. Klauder (Freeman, San Francisco, 1972), pp. 231–258
71. J. Alvarez-Muniz, J.L. Feng, F. Halzen, T. Han, D. Hooper, Phys. Rev. D **65**, 124015 (2002). arXiv:hep-ph/0202081 [hep-ph]
72. K.J. Mack, N. Song, A.C. Vincent, JHEP **04**, 187 (2020). arXiv:1912.06656 [hep-ph]
73. D.C. Dai, G. Starkman, D. Stojkovic, C. Issever, E. Rizvi, J. Tseng, Phys. Rev. D **77**, 076007 (2008). arXiv:0711.3012 [hep-ph]
74. P. Jain, D.W. McKay, S. Panda, J.P. Ralston, Phys. Lett. B **484**, 267–274 (2000). arXiv:hep-ph/0001031 [hep-ph]
75. M. Kachelriess, M. Plumacher, Phys. Rev. D **62**, 103006 (2000). arXiv:astro-ph/0005309 [astro-ph]
76. I. Romero, O.A. Sampayo, JHEP **05**, 111 (2009). arXiv:0906.5245 [hep-ph]
77. S. Pandey, S. Karmakar, S. Rakshit, JHEP **11**, 046 (2019). arXiv:1907.07700 [hep-ph]
78. N.S. Manton, Phys. Rev. D **28**, 2019 (1983)
79. F.R. Klinkhamer, N.S. Manton, Phys. Rev. D **30**, 2212 (1984)
80. S.H.H. Tye, S.S.C. Wong, Phys. Rev. D **92**(4), 045005 (2015). arXiv:1505.03690 [hep-th]
81. J. Ellis, K. Sakurai, JHEP **04**, 086 (2016). arXiv:1601.03654 [hep-ph]
82. J. Ellis, K. Sakurai, M. Spannowsky, JHEP **05**, 085 (2016). arXiv:1603.06573 [hep-ph]
83. S.F. Ge, M. Lindner, W. Rodejohann, Phys. Lett. B **772**, 164–168 (2017). arXiv:1702.02617 [hep-ph]
84. P. Coloma, P.A.N. Machado, I. Martinez-Soler, I.M. Shoemaker, Phys. Rev. Lett. **119**(20), 201804 (2017). arXiv:1707.08573 [hep-ph]
85. M.G. Aartsen et al. [IceCube], Nature **551**, 596–600 (2017). arXiv:1711.08119 [hep-ex]
86. S. Hummer, M. Maltoni, W. Winter, C. Yaguna, Astropart. Phys. **34**, 205–224 (2010). arXiv:1007.0006 [astro-ph.HE]
87. P. Mehta, W. Winter, JCAP **03**, 041 (2011). arXiv:1101.2673 [hep-ph]
88. S. Karmakar, S. Pandey, S. Rakshit, JHEP **10**, 004 (2021). arXiv:2010.07336 [hep-ph]
89. C.A. Argüelles, T. Katori, J. Salvado, Phys. Rev. Lett. **115**, 161303 (2015). arXiv:1506.02043 [hep-ph]
90. S.T. Scully, F.W. Stecker, Astropart. Phys. **34**, 575–580 (2011). arXiv:1008.4034 [astro-ph.CO]
91. J. Ellis, N.E. Mavromatos, A.S. Sakharov, E.K. Sarkisyan-Grinbaum, Phys. Lett. B **789**, 352–355 (2019). arXiv:1807.05155 [astro-ph.HE]
92. R. Laha, Phys. Rev. D **100**(10), 103002 (2019). arXiv:1807.05621 [astro-ph.HE]
93. V.A. Kostelecky, M. Mewes, Phys. Rev. D **70**, 031902 (2004). arXiv:hep-ph/0308300 [hep-ph]

Chapter 2
Sources of VHE (TeV-PeV) Neutrinos

The Earth is bombarded with cosmic-rays (CRs) from all directions isotropically. These energetic particles are believed to be accelerated in various galactic and extra-galactic sources. The flux of these CRs are extended over ~ 12 orders of energies starting from 10^9 eV [1]. But the sources of these energetic particles are still unknown. Being charged these particles are deflected in their path by galactic or extra-galactic magnetic fields, as a result they never point back to their sources.

CRs produce VHE γ-rays and neutrinos when they interact with matter or photon field inside the source or in the vicinity of the source. These high energy photons and neutrinos, being neutral, travel in a straight line. Therefore we can identify CR sources by detecting these photons and/or neutrinos. The γ-rays can interact with CMB or EBL photons along its path and get attenuated. Neutrinos, on the other hand, can travel unaffected because of their very small interaction cross-sections. For the same reason they can escape from dense interiors of astrophysical objects, which are opaque to electromagnetic radiation. Thus neutrinos are important messengers to identify CR sources and understand the underline physical processes that produce them.

2.1 Galactic Sources

It is generally believed that cosmic rays are accelerated up to energies 10^{15-18} eV in the Milky Way. Among various galactic sources supernova remnant (SNR), pulsar wind nebula (PWN) are the likely sources of these galactic cosmic rays.

© The Author(s), under exclusive license to Springer Nature Switzerland AG 2021
D. Bose and S. Rakshit, *High Energy Astrophysical Neutrinos*,
SpringerBriefs in Astronomy, https://doi.org/10.1007/978-3-030-91258-1_2

2.2   SNR and PWN

Supernova remnants are believed to be sources of cosmic-rays in the Milky Way. Stars burn huge amounts of nuclear fuel at their cores. Energy released in nuclear fusion processes prevents collapse of the star under its own gravity. For stars heavier that 8 solar masses, during the course of their evolution, heavier elements up to iron are formed. At this point the star no longer able to cancel out gravitational pull, core collapses under the gravitational force produced by its own mass known as Supernova (SN) Explosion. As the collapse happens heavy nuclei are dissociated by energetic photons and electrons are absorbed by protons producing neutrons via inverse beta decay. This leads to emission of enormous number of neutrinos of all flavours. Almost 99% of gravitational binding energy is taken away by the neutrinos. In 1987 around 24 neutrinos were detected from a Supernova explosion by three detectors Kamiokande, Baskan and IMB [2, 3].

Soon after the explosion stellar envelop which propagates through the inter stellar medium (ISM) is known as SNR. Cosmic-rays are believed to be accelerated to very high energies in these SNRs. They provide necessary power to maintain observed density of cosmic-rays in our galaxy. The energy Density of CRs in our Galaxy 1 eV/cm^3. In order to maintain this density, a source with acceleration power 10^{40} erg/s is needed. Average energy released in a SN explosion $\sim E_{SN} = 10^{51}$ ergs. If 0.01% of this energy transferred to CR particles and if there is 1 SN every 30 years then the acceleration power P = 0.01 $E_{SN}/30 = 10^{40}$ ergs/s i.e. the power required to maintain CR energy density.

Compact remnant of a core collapse supernova becomes either a neutron star (NS) or a black hole (BH). Some NS become pulsars. Pulsars are rotating NS. Shock wave which propagates outward carries away a fraction of the binding energy released and emits electromagnetic radiation. If a SNR has a pulsar at the core surrounded by pulsar wind called Pulsar Wind Nebula (PWN). Pulsars eject relativistic wind mainly composed of electrons, positrons. According to some models protons and ions are also present in the wind. The wind terminate in a standing wave and transfer a part of it's energy to accelerate particles. Neutrinos are produced when relativistic protons present in the wind interact with the surrounding matter.

2.3   Extra-Galactic Sources

The CRs with energies beyond 10^{18} eV are definitely of extra-galactic origin. The Larmor radius (R_L) (or gyroradius) of such energetic CRs are bigger than the galactic radius. They are likely to be produced inside the jets of (Active Galactic Nuclei (AGN), Gamma-ray Bursts (GRBs) and Star forming galaxies. Energy density of extra-galactic CRs is 3×10^{-19} erg cm^{-3}. The power required to maintain this energy density over the Hubble time scale i.e. 10^{10} years is 10^{-36} erg (Mpc)$^{-1}$ s^{-1}. It seems both AGN and GRBs can independently provide necessary power to maintain this energy density. For example, a GRB on average releases 10^{52} erg, rate of GRB is $300/(Gpc)^3$/year; if these GRBs convert a fraction of their energy to accelerate cosmic ray particles they can maintain the observed cosmic-ray density over Hubble time period. Similarly, AGN with number density 10^{-7} $(Mpc)^{-3}$ and bolometric luminosity 10^{44} erg s^{-1} can also provide required power to satisfy this condition.

2.4 Active Galactic Nuclei (AGN)

Active Galactic Nuclei or AGNs are a sub-class of galaxies whose core luminosity exceeds the total stellar emission. AGNs are powered by the gravitational energy released in the accretion processes by supermassive black holes (SMBH) $(10^{8-10} M_\odot)$. About 10% of all AGNs are more luminous at radio wavelengths than at optical ones and are, hence, called radio-loud. The radio emission is believed to originate in the associated jets of the spinning black hole. It is believed that particles undergo relativistic acceleration in those jets. The bolometric luminosity of AGNs are very high ($L_{bol} \simeq 10^{43-48}$ erg s^{-1}), therefore they are the ideal laboratories to study our Universe under relativistic conditions. In addition to the great energy output, they are also highly variable. This rapid fluctuation places strict limits on the maximum size of the energy source, because an object cannot vary in brightness faster than it takes light to travel from one side of its energy-producing region to the other.

At the centre of the AGN there is a SMBH which is surrounded by an accretion disk. Relativistic jets are located perpendicular to the plane of the accretion disk. Depending on the viewing angle of the jet with respect to the observer's line of sight, these objects exhibit different properties. Since particles are accelerated to relativistic energies inside the jet, therefore AGN with their jets oriented towards us are of utmost importance in high energy regime, known as blazars.

Blazars are a class of radio-loud AGN, whose one of the relativistic jets point close to our line of sight. Blazars are highly variable sources, variability ranges from few minutes to months. Because of narrow viewing angle all emissions are Doppler boosted in frequency by Doppler factor δ. The Doppler factor δ, of an object moving at very high speed v making an angle θ with the line of sight is given by

$$\delta = [\Gamma(1 - \beta \cos \theta)]^{-1} \tag{2.1}$$

where Γ is bulk Lorentz factor, defined by

$$\Gamma = \frac{1}{\sqrt{(1 - \beta^2)}} \tag{2.2}$$

The broadband continuum Spectral Energy Distributions (SEDs) of blazars are characterised by double hump structures. First hump, from radio through optical, UV, or even X-ray energies, is generally believed to be due to synchrotron emission by relativistic electrons present in the jet due to the magnetic fields associated with the jet. Second hump is extended from X-ray energies to γ-ray energies. Both leptonic and hadronic models are being proposed to explain the second hump. The second hump can be produced in two ways. Firstly, synchrotron photons are boosted to high energies via inverse Compton scattering by same electrons producing them, known as Synchroton-Self Compton (SSC) model. There is another possibility that relativistic electrons collide with photons produced outside the jet, known as External Compton (EC) model.

If high energy photons are produced by leptonic processes in that case it is difficult to confirm blazars as cosmic-ray sources. Therefore it is generally believed that the jets also contain relativistic protons. The high energy gamma-rays can be produced via synchrotron emission by relativistic protons or through photo-pion production when protons interact with radiation field or with other protons inside the emission region. Neutral pions will decay and produce high energy γ-rays and charged pions will produce muons, electrons, positrons and neutrinos via decay. Detection of TeV-PeV neutrinos will conclusively prove that cosmic rays are produced inside AGNs.

2.5 Gamma Ray Bursts (GRBs)

GRBs are the most powerful explosions in the Universe since Big Bang. These explosions release short bursts of γ-rays which lasts from milliseconds to thousands of seconds, known as prompt emission. These are catastrophic stellar events. The gamma-ray luminosity of these GRBs are in the range $10^{(51-53)}$ ergs/s, this is equivalent to the total energy released by the Sun during its entire lifetime in less than one second. GRBs were accidentally detected by Vela military satellites in 1967, and first results were published in 1973 [4]. In 1991, NASA launched BATSE (Burst and Transient Spectroscopy Experiment) onboard CGRO (Compton Gamma Ray Observatory) [1] to study these GRBs. It has detected 2704 GRBs during it's lifetime, which are distributed isotropically in the sky, indicating that they are of extra-galactic origin. The prompt emission is followed by a long afterglow emission, detected across electromagnetic spectrum from radio frequencies to VHE γ-rays. From the afterglow-observation key properties of GRBs such as nature of the progenitor and precise location can be determined.

Based on the duration of γ-ray emission during prompt phase, GRBs are classified into two groups, short and long GRBs. The short GRBs lasts less than 2s and if the prompt emission lasts for more than 2s then they are known as long GRBs. Afterglow emission of long GRBs revealed that there is presence of spectral line absorption of the continuum spectrum due to heavy elements in the host galaxy medium. Meaning long GRBs are associated with the death of massive stars or core-collapse supernova. Once the core of a massive star collapses, it leads to a black hole (BH). An accretion disk is formed around the BH. Relativistic jets are formed perpendicular to the

---

[1] https://heasarc.gsfc.nasa.gov/docs/cgro/cgro.html.

accretion disk. On the other hand SGRBs are not associated with any supernova events. They are found in elliptical galaxies, where there is very little evidence for star formation. The SGRBs are believed to be due to merger of two compact objects, two neutron stars or one neutron star and a black hole. In both NS-NS and NS-BH mergers a BH is formed surrounded by accretion disk and relativistic jets.

The enormous energy release by GRBs in such short times from very compact regions implies that produced luminosity exceeds Eddington luminosity and flings out matter in the explosion region, which gets heated up into a fireball of electrons, positrons, gamma-rays and also baryons. The fireball then expands relativistically. The matter is ejected in shells successively. At some point, the outer shells slow down and are caught by inner shells, internal shocks are produced by the collisions of these shells. The Synchrotron radiation produced by electrons escape at this phase, observed as prompt-emission. External shocks are produced when shells collide with interstellar medium, leading to afterglow emission.

2.5.1 Neutrino Production in GRBs

In a GRB neutrinos are produced at different stages with varying energies. The proton-photon ($p\gamma$) interactions are most likely source of neutrino production in GRBs. In comparison pp/pn interactions can take place at an early stage of a GRB when compactness is much higher at smaller radii e.g. when the jet is still inside the star. Charged and neutral pions are produced when a proton interacts with a photon. Neutral pions will produce gamma-rays and charged pions will produce neutrinos via decay.

GRBs can accelerate protons to energies greater than 10^{16} eV. These protons can interact with γ-ray photons (100 s of keV–MeV) emitted during the prompt phase and produce **PeV** (10^{15} eV) neutrinos. Accelerated protons can interact with with x-ray or optical/IR photons in the external shocks and produce **EeV** ($10^{16}-10^{19}$ eV) neutrinos. **TeV** (10^{12} eV) neutrinos are produced when protons interact with thermal photons trapped in the jet. It is possible that jet never breaks out of the stellar envelope, these are known as choked GRBs. Even though no photons can escape from core-collapsing stars due to very high optical depth, TeV neutrinos can still escape from those failed GRBs due to their low interaction cross-section. GRBs produce GeV neutrinos via nuclear inelastic collisions inside the relativistic jets. The GRB central engine generates thermal or MeV neutrinos, these MeV neutrinos are mostly associated with core-collapse supernova.

2.6 Star Bursts Galaxies (SBGs)

Starburst Galaxies are galaxies that are observed to be forming massive stars at an unusually fast rate compare to any normal galaxy like our Milky Way. It is believed

that this starburst activity is triggered by either galaxy mergers or by interaction
with galaxies nearby. Such interactions disturbs the dynamical equilibrium of the
interstellar gas. These massive stars have short lifetime. Observed optical and infrared
(IR) luminosity of SBGs imply very high densities of gas and of ambient radiation
fields, emitted by numerous young massive stars residing in those galaxies which
later explode as supernovae. Core-collapse supernovae (SNe) are therefore expected
to enrich the central star forming regions with relativistic protons and electrons, i.e.
cosmic-rays. Given the high density of target material, p–p interactions are highly
probable and will produce neutral and charged pions. Neutral pions will decay and
produce very high energy γ-rays and decay of charged pions, would then convert
part of the proton energy to neutrinos. Thus these SBGs are prime targets for VHE
and UHE neutrinos and VHE γ-rays and ideal objects to study the physics of CRs
and their impact on the ISM.

2.6.1   Tidal Disruptive Events (TDEs)

A super massive black hole (SMBH) is residing at the center of all galaxies. A
tidal disruptive event (TDE) happens when a star located at a large distance from a
SMBH is purturbed and start approaching it. As it comes nearer the tidal force of
the SMBH exceeds the stellar self gravity and as a result the star is ripped apart. The
subsequent accretion of stellar debris onto the black hole cause a flare. Such events
are particularly important for revealing presence of SMBHs which are mostly quiet,
unlike AGNs. TDEs also have relativistic jets where CR protons are accelerated via
shocks up to energy 10^{20} eV. Neutrinos can be produced either by $p\gamma$ or by pp
processes.

2.6.2   Glashow Resonance

In 1960 the theoretical physicist Sheldon Glashow had proposed that in the inter-
action of $\bar{\nu}_e$ with e^-, W^- boson will be produced [$\bar{\nu}_e + e \to W^-$] [5]. The cross-
secton ($\sigma(GR)$) for this Glashow Resonance (GR) process is enhanced for the energy
6.32 PeV in the electron rest frame.

These events are of utmost importance to identify astrophysical sources that
produce neutrinos via hadronuclear (pp) or photohadronic ($p\gamma$) interactions. The
astrophysical sources which accelerate CRs to relativistic energies, produce copious
amount of neutrinos through $p\gamma$ and pp collisions. Glashow resonance events can
be used as a discriminator of the relative abundance of the pp and $p\gamma$ processes. In
the $p\gamma$ processes mainly π^+ and π^0 are produced, whereas in pp processes all types
pions (π^+, π^-, π^0) are produced in equal numbers. As a result the relative content
of $\bar{\nu}_e$ from $p\gamma$ and pp collisions in the final state would be different.

In order to distinguish between $p\gamma$ and pp processes one needs to measure neutrino and anti-neutrino fluxes separately. IceCube or other current and upcoming telescopes which uses ice/water as medium to detect neutrinos can not separate neutrinos from anti-neutrinos by looking at the Cherenkov light patterns produced by them. This can be done by identifying the charged lepton produced in the cc interactions, using a magnetic field. A possible way out is to detect $\bar{\nu}_e$ flux by means of GR.

References

1. https://masterclass.icecube.wisc.edu/en/analyses/cosmic-ray-energy-spectrum
2. K. Hirata et al., Phys. Rev. Lett. **58**, 1490–1493 (1987)
3. R.M. Bionta et al., Phys. Rev. Lett. **58**, 1494 (1987)
4. Klebesadel et al., Astrophys. J. Lett. **182**, L85 (1973)
5. S.L. Glashow, Phys. Rev. **118**, 316–317 (1960)

Chapter 3
Detection of High Energy Neutrinos on the Earth

Neutrinos are the most weakly-interacting particles they can escape from dense interior of cosmic objects, which are opaque to electromagnetic waves. They are not deflected by magnetic fields in their path as they are charge-less. They point back to their sources of origin. At very high energies number of cosmic neutrinos arrive on Earth are very few. Some of these VHE neutrinos will interact with a nucleus and produce charged particles. These secondary particles will produce Cherenkov lights in a transparent medium. Therefore in order to detect these high energy neutrinos large detectors are needed, first proposed by Markov in 1970s [1, 2].

3.1 Ice and Water as Natural Mediums to Detect VHE Neutrinos

Naturally occurring large volume of ice and water can be used as medium to detect VHE neutrinos. IceCube neutrino telescope [3, 4] located in South Pole uses deep glacial ice and ANTARES [5, 6] located in the Mediterranean, uses deep sea water to detect neutrinos.

A neutrino when interacts in the ice or water produces secondary particles. These energetic particles travel in the medium with a speed more than the speed of light in the medium and thereby causes the medium to emit Cherenkov radiation [7–10], as shown in Fig. 3.1. This radiation is emitted in the wavelength range 300–400 nm i.e. in the optical-UV band of the electromagnetic spectrum. These photons can be detected by placing sensitive photosensors in ice or water. Photomultipliers or PMTs are the ideal for this purpose. A PMT converts Cherenkov photons into an amplified electrical pulse. The Cherenkov emissions are coherent and are beamed in the forward direction. The Cherenkov angle (θ_c) can be defined as,

© The Author(s), under exclusive license to Springer Nature Switzerland AG 2021   47
D. Bose and S. Rakshit, *High Energy Astrophysical Neutrinos*,
SpringerBriefs in Astronomy, https://doi.org/10.1007/978-3-030-91258-1_3

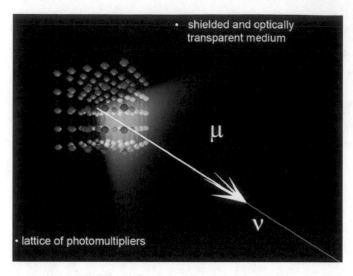

Fig. 3.1 A neutrino when collides with a nucleus in water/ice, produces charged particles. These particles emit Cherenkov radiation. Photo sensors immersed in the medium can detect the neutrino indirectly by collecting these photons. This figure is reproduced with permission from [11]

$$\cos(\theta_c) = \frac{1}{\beta \cdot n(\lambda)} \tag{3.1}$$

where β is equal to v/c and $n(\lambda)$ is the refractive index of the medium, which depends on the wavelength. For water and ice Cherenkov angle is $\sim 41°$.

Light has longer absorption length and shorter scattering length in the ice and other way round in the water. Therefore in ice detectors can be placed at larger distances compare to water. However due to smaller scattering lengths, arrival angle of the neutrino can be determined with high accuracy for experiments which uses water as the medium.

3.1.1 Neutrino Interactions in Ice and Water

At very high energies neutrinos interact mostly via deep inelastic scattering (DIS). There are two possible ways a neutrino can interact with matter, either by charge current interactions or by neutral current interactions. At 6.3 PeV an anti-electron neutrino can interact with a bound electron and produce W boson, known as Glashow resonance.

3.1.1.1 Charge Current (cc) Interactions of Neutrinos

The neutrinos when interact via cc interactions the charged lepton produced in that interaction carry almost 80% of the energy, rest of the energy is used to break the nucleus which then produces a hadronic shower.

In case of ν_μ (also $\bar{\nu}_\mu$) muons can travel several kilo meters in the medium before they decay. Because of long range muons produced outside the detector volume can enter the detector and leave a track like signature, as shown in [1] in Fig. 3.2. Muons will emit Cherenkov radiation along its path [12, 13]. At very high energies secondary muon is almost collinear w.r.t. neutrino [14]. The angle between μ and ν_μ can be parameterised as,

$$\Theta \approx 0.7° \left(\frac{E_\nu}{\text{TeV}} \right)^{0.6} \tag{3.2}$$

As a result angular resolutions for track like events are very good. However since muon deposits only a fraction of its energy in the detector energy estimation is not good for these events.

When ν_e (and $\bar{\nu}_e$) interacts in cc mode, electron (positron) will be produced, which will immediately initiate an electromagnetic shower. This electromagnetic shower will overlap with the hadronic shower produced at the vertex point. In detector topology such events will appear as spherical events as shown in [2] in Fig. 3.2. Such events are generally referred as shower events or cascade events.

The τ leptons have very short lifetime ($\sim 10^{-13}$ s), as a result they decay very soon. Most of the cases (65%) they decay hadronically and for rest of the time leptonically.

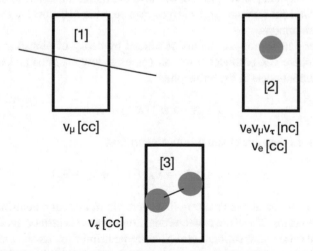

Fig. 3.2 Events in Detector topology: Tracks will be produced for $\nu_\mu(cc)$ interactions [1]. Neutral current interactions for all kind of neutrinos and charge current interactions of $\nu_e(cc)$ and $\nu_\tau(cc)$ will produce cascade events, which will appear spherical for a detector of km^3 size [2]. At energies > few PeV $\nu_\tau(cc)$ will produce dumble shaped "double bang" event [3]

Therefore at low energies ν_τs produce cascade like events in the detector. However as energy increases decay length of the tau increases (50 m at 1 PeV or 500 m at 10 PeV), which results in semi-track like events [15]. If the first interaction point and tau decay point lie inside the detector then the event will look like a double bang (two showers separated by a short distance) as shown in [3] in Fig. 3.2. And if the second shower i.e. only the tau decay point lie inside the detector then the event will appear as lollypop. Such events can easily be distinguished from a muon track event as the later will not produce a shower.

3.1.1.2   Neutral Current (nc) Interactions of Neutrinos

In the nc interaction neutrino only transfer a fraction of energy to break the nucleus and moves away with remaining energy. Therefore for such interactions only hadronic showers will be produced at the vertex point i.e. cascade events will be produced on the detector for neutral current events for all flavours. Cascade events has good energy resolution but moderate angular resolution.

3.1.2   Background Events: Atmospheric Muons and Atmospheric Neutrinos

The experiments which uses water/ice to catch astrophysical neutrinos, dominated by events induced by atmospheric muons and atmospheric neutrinos. Both are produced when cosmic-ray interacts with a molecule in the atmosphere. The muons can penetrate very deep underground, because of this reason neutrino detectors are usually built well beneath the surface. Even then neutrino detectors are overwhelmed by atmospheric muons.

Atmospheric neutrinos are mainly produced by decay of pions/kaons [16–19]. In the atmosphere CR produces a meson (pions at low energies or kaons at high energies) when interacts in the atmosphere,

$$p + N \rightarrow \pi^+(K^+) + X \tag{3.3}$$

These mesons then decay and produce neutrinos,

$$\pi^+ \rightarrow \nu_\mu + \mu^+ \rightarrow \nu_\mu + (e^+ + \nu_e + \bar{\nu}_\mu) \tag{3.4}$$

It is clear from the above equation that there are more muon neutrinos compare to electron neutrinos. These are conventional atmospheric neutrinos. IceCube experiment has measured conventional atmospheric neutrino flux, as shown in Fig. 3.3. The spectrum of conventional atmospheric neutrinos ($\propto E^{-3.7}$) is one power steeper compare to the spectrum of cosmic-rays ($\propto E^{-2.7}$).

Atmospheric neutrinos are also expected to be produced by decay of heavy mesons, which contains charm quarks. Prompt atmospheric neutrinos would have same power law spectrum as that of cosmic-rays.

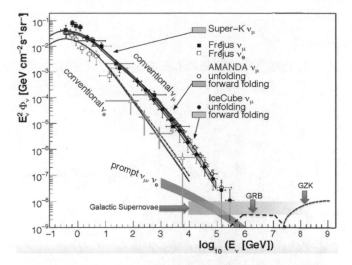

Fig. 3.3 Conventional atmospheric neutrino flux (for ν_μ and ν_e) shown here measured by IceCube telescope. Expected flux of prompt component shown by the magenta curve. Neutrino flux of cosmic origin will dominate at higher energies. This figure is reproduced with permission from [11]

Astrophysical neutrinos are generally believed to be produced by cosmic-rays, which are accelerated by Fermi acceleration mechanism thus will have power-law spectrum of the form $\phi \propto E^{-2}$. Therefore cosmic neutrinos will dominate above 100 TeV (Fig. 3.3).

3.1.3 IceCube

IceCube neutrino observatory located at the geographical south pole in the Antarctica is the largest underground detector to catch astrophysical neutrinos. It uses 1 km^3 of dark and highly transparent glacial ice, weighing a gigaton, to detect neutrinos (Fig. 3.4). Thousands of PMTs attached in strings observe for Cherenkov emissions produced by secondary charged particles, created when a neutrino interacts with an ice molecule. There are total 86 strings deployed at regular intervals. Each of these strings contains 60 digital optical modules (DOMs). A DOM is a basic detection element of the IceCube detector, a complete data acquisition system in a 35 cm spherical glass sphere. It is consists of a 25 cm diameter PMT at the bottom facing downwards and digitization electronics and power supply in the upper part, shown in Fig. 3.5. There are total 5160 such DOMs. A DOM digitises the Cherenkov pulse produced by the PMT, puts a time stamp and sends the digitised signal to the data acquisition room at the surface. These DOMs are located at the depth between 1.45 and 2.45 km below the surface. Horizontal spacing's between any two string is 125 m and vertical spacings between DOMs is 17 m.

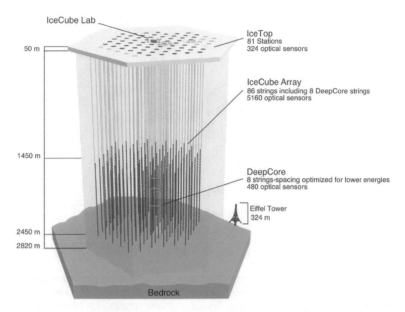

Fig. 3.4 Schematic of IceCube neutrino telescope at the South pole. DOMs are located at the depth from 1450 to 2450 m in the Antarctic ice. Total instrumented volume 1 km^3. *Credit* IceCube Collaboration

Fig. 3.5 Photograph of a DOM. *Credit* Robert Schwarz, NSF

The core of the array is denser with 8 strings spaced with smaller horizontal spacing's between them (70 m), known as DeepCore. This helps to reduce the energy threshold to ∼10 GeV. Without DeepCore, IceCube has energy threshold around 100 GeV.

There are two tanks, with DOMs inside, at the top of each string on the surface. This surface array, known as IceTop, is used to measure extensive air showers produced by cosmic-rays in the atmosphere. These tanks are also used a veto to reject atmospheric muons entering the inner detector.

3.1.3.1 IceCube Events

Neutrinos interacts via weak interaction, therefore most of them will simply go through the detector without leaving any trace. Once in a while a neutrino will interact with ice and produce secondary charged particles, these particles will emit Cherenkvo radiation. Arrival direction of the incoming neutrino will be reconstructed from the arrival times of Cherenkov photons at different DOMs and energy of the neutrino can be reconstructed from the total number of photons detected. As explained earlier Ice-Cube will see either track-like events or cascade-like events. The track-like events are produced by through-going muons produced in $\nu_\mu(cc)$ interaction, shown in Fig. 3.6. Very high energy track events has sub-degree resolution (0.2°–0.3°), therefore ideal for neutrino astronomy.

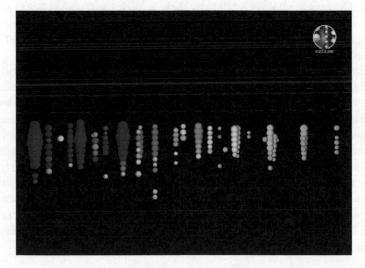

Fig. 3.6 Shown here is a reconstructed track event for IceCube. A muon produced in $\nu_\mu(cc)$ interaction travel through the detector emitting Cherenkov radiation. DOMs indicated with red will receive light earlier and DOMs indicated with blue will detect light later in it's path. Using this delay in arrival times, arrival direction of the muon can be estimated. *Credit* IceCube Collaboration

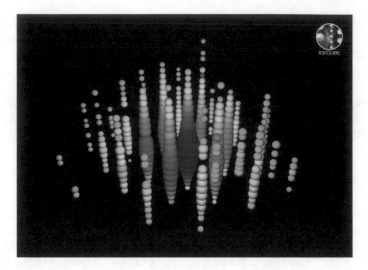

Fig. 3.7 A reconstructed cascade event. For cascade events a hadronic shower will be induced by the neutrino inside the detector volume. Light from this shower will then spread out in all directions. DOMs indicated by red will see light first and DOMs indicated by blue will see light later. *Credit* IceCube Collaboration

The cascade-like events are produced by ν_e and ν_τ (cc) interactions and nc interactions of all flavours (Fig. 3.7). Due to multiple scatterings of secondary charged particles, Cherenkov photons emitted by these particles will spread out in all directions, as if emanating from a point source. Thus angular resolution of these events are not good. However since most of these events will be well contained energy reconstruction can be done with high precision (10–15%).

As explained earlier, IceCube events are hugely dominated by atmospheric muons and atmospheric neutrinos [16, 17]. Astrophysical neutrinos can be identified against this overwhelming background in two ways. Firstly, IceCube can look for thoroughgoing muon tracks produced by upcoming muon neutrinos. In this method detector is sensitive only to half of the sky, only for sources located in the northern hemisphere. In the second method, DOMs located in the outer strings act as veto, atmospheric muons entering the detector from outside will be tagged and hence can be rejected. IceCube is sensitive for all sky and for all favours in this method. However total fiducial volume (∼500 megaton) used for detecting cosmic neutrinos is reduced.

The number of events expected for an astrophysical source can be obtained as follows. Firstly, one needs to estimate the probability $P(E_\nu)$ of interaction for a neutrino, which is given by,

$$P(E_\nu) = 1 - \exp\left[-\left(\frac{L}{\lambda_\nu(E_\nu)}\right)\right] \simeq \frac{L}{\lambda_\nu(E_\nu)} \qquad (3.5)$$

where L is the path length traversed inside the detector volume by the neutrino and λ_v is the mean free path in the ice.

$$\lambda_v = [\rho_{ice} N_A \sigma_{vN}(E_v)]^{-1} \tag{3.6}$$

where ρ_{ice} (=0.9 g cm^{-3}) is the density of the ice, N_A (=6.023×10^{23}) is Avogadro number and σ_{vN} is the neutrino-nucleon cross-section.

Once probability is known then number of events can be obtained in a given time t as,

$$N_{evts} v = t \int A_{eff}(E_v) P(E_v) \phi_v dE_v \tag{3.7}$$

Neutrino interaction cross section increases with energy. At energies above 100 TeV the earth becomes opaque to the highest energy neutrinos. As a result, IceCube can only observe neutrinos from the horizon or above.

3.1.4 IceCube Highlights

IceCube for the first time in the year 2013 reported to have detected two very high energy astrophysical neutrino events [20] in two years. Measured energy of these two events were 1.04 and 1.14 PeV. These events were found to be incompatible with background atmospheric neutrinos with 2.8σ. Both events were cascade like and fully contained, thus energy was estimated with high precision. These events were induced either by nc interaction of neutrino of any flavour or by cc interaction of v_e or $\bar{v}_e$. Later in that same year IceCube reported to have detected 26 additional neutrino events of astrophysical origin [18]. All these events were discovered in an all sky search. In order to suppress the overwhelming background, DOMs located in the boundary are used as active veto, events which are started within the detector volume are accepted thus known as high energy starting events (HESE). So far IceCube has detected 102 such HESE events in 7.5 years above 10 TeV [21]. Most of these events are cascade like, which is expected for a sample of astrophysical neutrinos, as these neutrinos are expected to have equal flux for each flavour. IceCube has also found evidence for astrophysical neutrinos in other channels independently: through-going tracks [22–24] and cascade like events [25]. Spectral index measured for cascade channel is −2.53 and that for high energy muon tracks originated from Northern hemisphere is −2.2. The best fit power-law index to the combined data set i.e. tracks and cascade is −2.5 [26]. IceCube detects approximately 30 VHE astrophysical neutrinos per year. Majority of these events are distributed isotropically in the sky, meaning they are of extra-galactic origin i.e. produced by AGN. GRBs or SBGs. So far only one event is identified with a source. In the year 2017 IceCube for the first observed one $v_\mu(cc)$ event, IC-170922A, which was coincident in direction with an active galactic nuclei TXS 0506+056. Interactions which produces neutrinos also produces gamma-rays. Dedicated search for correlation among detected gamma-ray sources revealed that only 30% of these astrophysical neutrinos could be coming from AGN [27] and contribution from GRBs not more than 1%. This imply that majority of the cosmic neutrino events detected by IceCube come from sources which are opaque to gamma-rays. IceCube has also identified two ν_τ events with 2.8σ significance [28].

3.1.4.1   TXS 0506+056 Event

On 22nd September, 2017 IceCube detected neutrino track event (IceCube-170922A) (a ν_μ CC event) with energy >290 TeV,[1] which was coincident in direction with a blazar TXS 0506+056 [29]. This blazar is located at distance 5.7 billion light-years (redshift z = 0.3365) from us. This is for the first time a high energy neutrino event is detected in association with an astrophysical object outside our galaxy. The other extra-galactic source previously observed was supernova 1987a at much lower neutrino energies. Within one minute of the neutrino detection, IceCube sent an automated alert to astronomers around the world and subsequently a multi-wavelength campaign, including radio, infrared, optical, X-rays and gamma-rays, was followed involving telescopes across the globe. This source was in a flaring state at that time. Further analysis of archival data from the direction of TXS 0506+056 revealed a "neutrino flare" above atmospheric background between September 2014 and March 2015. A total 13 ± 5 neutrinos were detected with a significance of $\sim3.5~\sigma$ [30]. However unlike IceCube-170922A event, this past neutrino flare was not accompanied by a gamma-ray flare or high flux in any other wavelength. Detection of high energy neutrinos implies that hadrons are accelerated inside the jets of this object and therefore TXS 0506+056 is a definite source of cosmic-rays.

3.1.4.2   TED Event

IceCube has detected one TDE event on 1st October, 2019. Based on measured energy (0.2 PeV) this event was estimated to be of astrophysical origin with 59% probability [31]. Following IceCube alert Zwicky Transient Facility (ZTF) few hours later found a TDE source $AT2019dsg$ in that direction which was identified as a candidate neutrino source. A neutrino typically carries 0.05 of the proton energy, meaning parent proton should have at least 4 PeV energy. Neutrinos can be produced either by $p\gamma$ or by pp processes . If UV photons from photosphere are the target for photohadron interactions then estimated neutrino energy (E_ν) is ≈0.8 PeV. And if x-rays are target photons then expected E_ν is ≈0.05 PeV. Both these values are compatible with measured energy 0.2 PeV, taking into account systematic uncertainties. This result also indicate that at least 3% of the observed diffused astrophysical neutrinos seen by IceCube are from TDEs.

---

[1] https://gcn.gsfc.nasa.gov/gcn3/21916.gcn3.

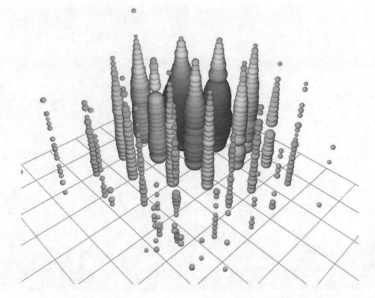

Fig. 3.8 A visualization of the Glashow event as seen by the IceCube detector. Each colored circle represents a DOM which was triggered by the Cherenkov photons produced by this $\bar{\nu}_e$ event. Red circles indicate sensors triggered earlier in time, and green-blue circles indicate sensors triggered later. Size of the circles indicate number of photons detected by the DOM. *Credit* IceCube Collaboration

3.1.4.3 Glashow Resonance

IceCube has detected one Glashow Resonance event on 8th December, 2016 with 5σ significance [32]. This event was found during an archival analysis using machine learning based algorithm with a sample of PeV energy events which are partially contained in the detector volume. Measured energy for this event was, after taking into account systematic uncertainties, 6.05 ± 0.72 PeV, as expected for a GR event. Further analysis revealed that this event can be fitted with a hadronic shower and a muon track emanating from the shower, indicating that the W^- boson decayed hadronically (Fig. 3.8).

3.1.5 ANTARES

The ANTARES (Astronomy with a Neutrino Telescope and Abyss environmental RESearch) is a water based neutrino located at a depth of 2475 m in the Mediterranean Sea 40 km off shore from Toulon, France. There are total 12 strings with average spacings 60–70 between them [33]. Total instrumented volume 0.1 km^3. There are almost 900 optical modules (OM), consists of PMTs, collects Cherenkov lights produced by charged secondaries. Since light has small scattering length in water,

Fig. 3.9 View of the ANTARES detector. *Credit* J. A. Aguilar (2010) for the ANTARES Collaboration

therefore ANTARES has better angular resolution compare to IceCube experiment. ANTARES has high sensitivity for sources in the galactic plane and for extra-galactic sources in the southern sky [34, 35] (Fig. 3.9).

ANTARES due it's location, even with much smaller volume compared to Ice-Cube, can study the Southern sky and the Galactic centre with very high significance using upward going muon events. Also, since it uses water as medium it has better angular resolution even for cascade like events. However to date, no counterpart has been identified for any neutrino event. They have given best limit for galactic neutrino emission [36]. Jointly with IceCube they have searched for neutrino sources in the Southern sky with best sensitivity ever, but found no significant evidence for any sources [37]. Similarly they were part of many multi-messenger campaigns to seek connection between neutrinos and other cosmic messengers, they followed-up gravitational wave events [38, 39], the neutrino source TXS 0506+056 [40], studied blazars detected by Fermi-LAT and ground based gamma-ray telescopes which were in flaring state between 2008 and 2012 [41].

3.1.6 Upcoming Neutrino Detectors

IceCube neutrino telescope has revolutionized field of astro-particle physics by detected neutrinos of astrophysical origin. However it is now clear that bigger detectors are needed to identify cosmic neutrino sources. Also, at higher energies for full sky coverage detectors are required at both hemisphere, since as energy increases the earth becomes opaque for these neutrinos.

Currently two detectors are under construction in the northern hemisphere KM3NeT and GVD. The IceCube collaboration also plans to build a bigger detector **IceCube-gen2**, in order to identify neutrino sources with high significance.

3.1.6.1 IceCube-Gen2

The next generation IceCube-Gen2 [42, 43] is the planned upgrade for IceCube neutrino telescope. It will have ($8 km^3$) in-ice optical array. For the optical array another 120 strings will be added to the existing array with horizontal spacings 240 m. There will be a denser array at the core for low energy neutrinos Precision IceCube Next Generation Upgrade (PINGU) [44], a much wider surface array to detect air showers produced by CRs in the atmosphere and veto down-going penetrating muons present in the air showers. And a very large, $500 km^2$ radio array, for detection of UHE neutrinos. The expected sensitivity to identify a point source of IceCube-Gen2 would be 5 times better compare to present IceCube array. It is therefore will detect ten times more astrophysical neutrinos each year, in the energy range TeV to EeV, thus increasing chance probability to discover astrophysical neutrino sources. Main scientific goals of this next generation IceCube detector would be to reveal neutrino sources and understand acceleration mechanism through multi-messenger observations.

3.1.6.2 KM3NeT

The KM3NeT, a multi-cubic kilometer neutrino detector is currently under construction in the Mediterranean sea [45]. It will have two major components. **ARCA (Astroparticle Research with Cosmics in the Abyss)**, high energy component, will be located offshore Capo Passero in Sicily-Italy. There will be two arrays, each containing 115 lines. Each of these lines will have 18 DOMs. Total instrumented volume of ARCA will be 1 gigaton. ARCA will mainly look for neutrinos coming from galactic and extra-galactic sources. Due to its location primary goal of the ARCA is to find neutrinos from the CR accelerators in our Milky Way. Expected angular resolutions for track and cascade events would be 0.1° and 1.5° respectively. As a result estimated sensitivity of ARCA will be orders of magnitude better compare to IceCube for sources located in the Southern Hemisphere. Other component is **ORCA (Oscillation Research with Cosmics in the Abyss)**, low energy component will be located offshore Toulon, France. The main science goal for ORCA is to study oscillation physics. Both ARCA and ORCA will use multi-PMT DOMs to detect neutrinos. These DOMs have large photo-cathode coverage and better angular sensitivity w.r.t. single PMT DOMs.

3.1.6.3   Baikal-GVD and P-ONE

The Baikal-Gigaton Volume Detector (Baikal-GVD) is the neutrino detector under construction in the lake Baikal, Russia [46]. There will be total 14 cluster with 8 strings in each of them. Each of these strings will have multiple optical modules. Baikal-GVD will have $0.7\,km^3$ detection volume.

The Pacific Ocean Neutrino Explorer (P-ONE) detector will be built with the aim to detect high energy astrophysical neutrinos in the energy range $10\,TeV$–$10\,PeV$ [47]. The plan is to build a segmented array of several cubic kilometers of water in the Pacific ocean. This is a cost effective way to instrument a large volume of water, with fewer strings.

3.2   Detection of Ultra High Energy Neutrinos

The Ultra High Energy (UHE) neutrinos (PeV to EeV) are expected to produce when Ultra-High Energy Cosmic Rays (UHECRs) interact with CMB photons. However expected flux of these cosmogenic or GZK neutrinos is too small, $1\,neutrino/km^2/year/steradian$. Therefore very large detectors are needed to observe these neutrinos. Even the planned extension of the Optical array of IceCube-Gen2 is not big enough for detection of UHE neutrinos. A radio array will be added for that purpose.

There exists couple of techniques by which large detectors can be built in cost effective way to detect UHE neutrinos on Earth. Firstly, these neutrinos can be observed by detecting coherent radio emission produced by secondary charged particles present in the shower initiated by the neutrino in a dense medium, e.g. ice as well as in the atmosphere. Or by detecting charged secondaries in solid state detectors used for cosmic-ray-induced EAS.

In both techniques only down-going neutrinos and Earth skimming (neutrinos which incident at large inclined angles w.r.t. zenith angle) neutrinos can be observed.

3.2.1   Radio Emission by EM Component of the Cascade

There are two kind of radio emissions happen in the medium when a cascae is initiated in the medium by UHE neutrinos, radiation due to the net charge excess in the shower, known as Askaryan effect and radiation by charged particles due to the deflection in the geomagnetic field.

3.2.1.1   Askaryan Radio Emission

An UHE neutrino initiates a cascade in a dense medium via neutral current or charge current interactions. In the first case hadronic cascade produced at the vertex point is eventually turns into an electromagnetic cascade. For cc interactions the resulting shower is basically EM in nature. In the 1960s Gurgen Askaryan predicted that an electromagnetic cascade in a dielectric medium such as ice, would emit coherent Cherenkov like radiation in the radio wavelengths [48, 49].

An EM shower, consists of electrons, positrons and photons, is mainly developed by pair production and bremsstrahlung. Initially the shower is neutral as electrons and positrons are always produced in pairs and lose energy equally. Shower is developed in the longitudinal direction as long as an electron's rate of energy loss due to ionisation is equal to the rate of energy loss due to bremsstrahlung [50].

The charge asymmetry happens as photons produced via bremsstrahlung drag atomic electrons into the cascade by Compton scattering. This is the main contributor for excess negative charge. Also there are other interactions which add extra electrons into the shower, e.g. when positrons interact with atomic electrons via Bhabha scattering and via annihilation in the flight. Overall a $\sim$20% negative charge asymmetry is developed. The charge excess then emit Cherenkov like radiation.

If the wavelength of the emitted radiation is greater than lateral or transverse spread is characterised by Moliere radius (R_M) i.e. $\lambda > R_M$ then wavelengths will add coherently. In that case emitted power is proportional to square of the number of particles ($P \propto N^2$), radiation is emitted in the frequency range 100 MHz to 1 GHz. And if $\lambda < R_M$ then wavelengths will interfere destructively.

The angle of emission θ_c, is given by $\cos \theta_c = \frac{1}{n\beta}$, where n is the refractive index of the medium and β is the speed of the charge particle in terms of speed of light. For the glacial ice in the South Pole the angle of emission is $\theta_c \sim 56°$ since n is 1.76.

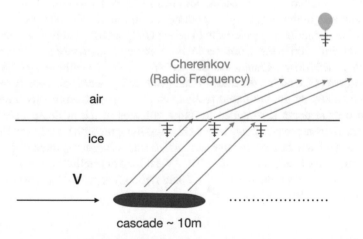

Fig. 3.10 Glacial ice at South Pole, glacial ice in Greenland and Moore's Bay on the Ross Ice Shelf are the ideal sites to detect UHE neutrinos. These neutrinos will initiate a EM shower in the ice. Cascade will develop over a short length $\sim$10 m. Charged particles in the cascade will emit Cherenkov emission in radio frequencies. Emission is vertically polarised. These radiations can be detected either by burying antennas inside (e.g. ARA, ARIANNA) or by flying detectors in a balloon (as done by ANITA experiment)

Therefore for neutrinos arriving at large angles w.r.t. zenith, only the top portion of the Cherenkov emission will be incident on the ice-air as shown in the Fig. 3.10, meaning Askaryan emission will be vertically polarised. The Askaryan effect was confirmed at SLAC National Accelerator Laboratory in 2001 in a beam test. Coherent radio emission were detected by microwave horn antennas [51]. Later similar tests was carried out on rock salt [52] and on ice [53]. It appears that ice is an ideal target for detection of UHE neutrinos because of it's radio transparency and long attenuation length ∼1 km.

3.2.1.2 Geomagnetic Radio Emission

An UHE neutrino can induce an EAS in the atmosphere. Electrons and positrons present in the EAS can produce geomagnetic radiation when they are deflected by the Earth's magnetic field. Each particle will experience the Lorentz force F,

$$\bar{F} = q(\bar{v} \times \bar{B}) \tag{3.8}$$

where v is the velocity of the particle and B is the Earth's magnetic field and q is the charge of the particle. Emitted radiation is observed below 100 MHz. Unlike Askaryan emission, geomagnetic emission is linearly polarized orthogonal to the direction of the geomagnetic field and shower axis. Therefore at South Pole this emission will be horizontally polarised. The geomagnetic radiation is proportional to longitudinal extension and since in dense medium cascade dimension is only few meters it is negligible in dense media, dominates only in air showers. The Askaryan effect dominates in the dense medium but not negligible for air showers.

Most of the experiments which aim to detect UHE neutrinos, rely on observations of Askaryan emission in the dense media. For this they need to monitor large volume of dielectric medium transparent to radio waves. ANITA is a balloon borne detector, which observes Glacial ice for radio signal produce by neutrinos. Askaryan signal can also detected by ground-based techniques that deploy instruments directly on the surface or at depth surrounded by large volumes of ice in South pole and in Greenland. There are experiments which observed lunar regolith from Earth. Putting a detector farther from the neutrino interactions increases the energy threshold. Surface detectors thus have energy threshold ∼10^{17} eV, balloon or satellite borne experiments have energy threshold around 10^{18-19} eV and radio antennas which observe Moon have energy threshold more than 10^{20} eV.

3.2.2 ANtarctic Impulsive Transient Antenna (ANITA)

The ANtarctic Impulsive Transient Antenna (ANITA) rely on detection of Askaryan emission by cascade induced by UHE neutrinos in the glacial ice at South Pole. It monitors large volume of ice from an altitude of 36–37 km above sea level (33–35 km

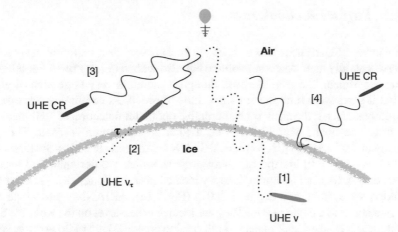

Fig. 3.11 ANITA experiment is sensitive to following events, [1] Askaryan radiation produced by cascade induced by UHE neutrinos in the ice. [2] Upcoming ν_τ first interacts in the ice, resultant τ can travel several tens of meters and then decays in the atmosphere initiating a cascade. This cascade will emit radio emission. [3] Showers initiated by UHE cosmic-rays at large zenith angle, radio emission from such showers will be detected directly by ANITA. [4] ANITA can also detect reflected radio emission from UHE cosmic-ray induced shower in the atmosphere. For showers in the atmosphere, main contributions will come from geomagnetic radiation but a small contribution will come from Askaryan emission (shown by solid lines). For cascades initiated in the ice will have only Askaryan radiation (shown by dotted line)

above the ice surface). It is consists of balloon-borne array of omni-directional radio antennas. It aims to catch neutrinos above $E_\nu > 10^{18}$ eV. It can observe 10^6 km^2 at any point of time and since attenuation length of radio waves is $\sim$ 1 km, meaning total interaction volume for ANITA is 10^6 km^3. ANITA can detect UHE neutrinos arriving at large angle (Earth-skimming) and upcoming ν_τs as depicted in Fig. 3.11. It can also detect radio signals produced by charged secondaries present in the EAS initiated by a cosmic-ray in the atmosphere (as shown in the Fig. 3.11).

The ANITA experiment has monitored Antarctic ice in three different missions, first one launched on December 15, 2006. ANITA-1's antennas are 32 dual-linear-polarization, quad-ridged horns, bandwidth 200–1200 MHz. Second mission was launched on December 21, 2008, with 40 horn antennas. ANITA-III was launched on December 18, 2014, it had 48 antennas and better sensitivity w.r.t. previous two missions.

ANITA experiment has detected few anomalous events during it's flights [54–56]. They are found during first and third flights. These events are appeared to be ν_τ events. The upward-going ν_τ can interact with earth via CC interaction and then the resulting τ will decay in the atmosphere and produce EAS. Impulsive radio pulses produced by charged particles in EAS can trigger ANITA detector. However these events, poses severe challenge for standard model interpretation of astrophysical origin. If these neutrinos are of astrophysical origin then they would be accompanied by several neutrinos (all flavours) in TeV-PeV range. IceCube neutrino telescope carried out searches for such events [57] but found none.

3.2.3   Surface Radio Arrays

The number of UHE neutrinos arriving on earth is very few, therefore one needs to observe not only huge detection volume but also with very long time. For a balloon borne experiment like ANITA, even though it monitors very large area of glacial ice but it's exposure time is very small little above 100 d combining all missions. Therefore surface radio array is an alternative option for detection of UHE neutrinos which can be operational 24 × 7 even if total coverage area is small. The ARA (Askaryan Radio Array) [58, 59] and ARIANNA (Antarctic Ross Ice-Shelf ANtenna Neutrino Array) [60] are the two pilot projects which are designed and tested at Antarctica. ARA is an in-situ radio array located near the South Pole. The antennas for ARA are situated at a depth of 200 m (Fig. 3.12). ARIANNA proto-type array has 10 stations at a depth of 2 m. They are located at sea-level on the Ross Ice Shelf, Antarctica. IceCube-Gen2 plans to built a wide area 500 km² radio array based on the experience gained from ARA.

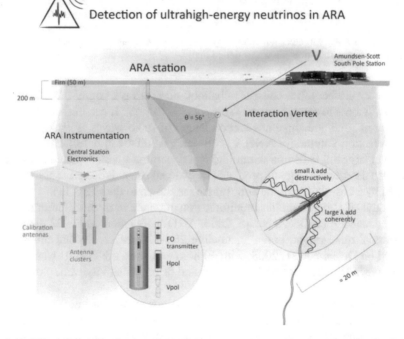

Fig. 3.12 The Askaryan Radio Array (ARA) is a proto-type neutrino detector at the South Pole. ARA is designed to detect and measure UHE cosmic via Askaryan radio emission. The detector elements are buried in the ice, about 200 m deep. *Credit* WIPAC/ARA

3.3 UHE Neutrino Detection via Moon

The lunar regolith, made of dust and rocks is also a very transparent medium for radio emission, thus can be used for UHE neutrinos above 10^{19} eV. The cascade induced by these neutrinos will generate Askaryan radio emission which can be detected by radio antennas on Earth. Multiple experiments in the past have observed Moon for UHE neutrinos, the Parkes Lunar Radio Cherenkov Experiment in 1965 [61], GLUE [62], LOFAR [63] etc. The upcoming Square Kilometer Array (SKA) [64] will be even more sensitive to neutrinos from the moon with this technique due to its wide bandwidth and large collection area.

3.4 Detection of UHE Neutrinos by AUGER Air Shower Array

As discussed earlier UHE neutrinos (all flavours) can initiate an EAS in the atmosphere. Secondary charged particles present in the shower will produce Cherenkov lights in the atmosphere or in water/ice along their path. The AUGER air shower experiment is designed to detect EAS produced by UHE cosmic rays in the atmosphere, can also detect UHE neutrinos. Initially EAS will have a large amount of electromagnetic component, as the shower grows later it is dominated by muons. Since neutrinos have very small interaction cross-section w.r.t. cosmic-rays, they are expected to interact deep inside the atmosphere. Therefore EAS generated by neutrinos arriving at very large angle w.r.t. zenith will retain electromagnetic component even at the observation level. On the other hand EAS produced by cosmic-rays will develop higher up in the atmosphere and therefore will be dominated by muons, they will leave different signature in the detector, which can be used to separate out cosmic-ray events from neutrino events. AUGER experiment located in Argetina uses it's surface detectors to observe UHE neutrinos. These surface detectors are tanks filled with pure water. A secondary charge particle will produce Cherenkov light in the water which is then collected by PMTs placed inside. There are total 1600 detectors distributed over 3000 km^2. The AUGER experiment searches for downgoing neutrino events from showers arriving at large angles (within zenith angles 75°–90°). AUGER also looks for Earth-skimming showers induced by tau neutrinos which interact in the Earth's crust (within zenith angles 90°–95°) [65].

AUGER collaboration searched for neutrino events—from highly-inclined air showers induced by neutrinos of all flavours and Earth-skimming tau-neutrinos from 2004 till 2018 and has not found any neutrino event [66]. They carried out a joint search for Binary Neutron Star Merger GW170817 with ANTARES and IceCube [38]. On August 17, 2017, the Advanced LIGO [67] and Advanced Virgo [68], GW detectors detected a GW signal, GW170817, from a binary neutron star inspiral. Soon Fermi-GBM and INTEGRAL detected a short GRB GRB170817A, consistent with GW170817. Pierre AUGER collaboration did a joint search for any neutrino event from this GW event along with IceCube and ANTARES neutrino observatories. The zenith angle for this source at the time of detection was 91.9°, falls in the FOV of AUGER for Earth-Skimming tau-neutrino events. However no coincident neutrino event was found.

3.5 Acoustic Detection

The UHE neutrinos can also be detected by acoustic method. The rapid heat deposited by the cascade induced by a neutrino in a dense medium will produce a pulse of sound in the frequency range 10–30 kHz [69, 70]. This phenomena has been observed experimentally [71]. The basic idea is that when a UHE neutrino interacts in ice or water initiates a hadronic shower, which typically carries 25% of the neutrino's energy. Cascade develops over a small cylindrical volume of radius $\sim$ few cms and length $\sim$10 m. Meaning huge amount of energy is deposited in a small volume. As a result the medium expands rapidly in the perpendicular direction. A bipolar pulse is produced. At present some proto-type experiments are undergoing in the Mediterranean sea and lake Baikal [72].

References

1. M.A. Markov, I.M. Zheleznykh, Nucl. Phys. **27**, 385–394 (1961)
2. A. Roberts, Rev. Mod. Phys. **64**, 259–312 (1992)
3. https://icecube.wisc.edu/wp-content/uploads/2020/11/IceCubeDesignDoc.pdf
4. J. Ahrens et al., Astropart. Phys. **20**, 507–532 (2004)
5. J. Aguilar et al., Astropart. Phys. **26**, 314–324 (2006)
6. E. Migneco, J. Phys. Conf. Ser. **136**, 022048 (2008)
7. J.D. Jackson, *Classical Electrodynamics* (Wiley, New York, 1975)
8. J.V. Jelly, *Cherenkov Radiation* (Pergamon, New York, 1958)
9. P.A. Cerenkov, Phys. Rev. **52**, 378–379 (1937)
10. I. Frank, I. Tamm, C. R. Acad. Sci. USSR **14**, 109–114 (1937)
11. M. Ahlers, F. Halzen, Rep. Prog. Phys. **78**, 126901 (2015)
12. http://pdg.lbl.gov/2013/AtomicNuclear
13. L. Radel, C. Wiebusch, Astropart. Phys. **38**, 53–67 (2012)
14. The ANTARES Collaboration, arXiv:astro-ph/9907432 (1999)
15. D. Chirkin, W. Rhode, hep-ph/0407075 (2004)
16. F. Halzen, S.R. Klein, Phys. Today **61**(5), 29 (2008)
17. F. Halzen, https://www.trisep.ca/2018/program/trisep_halzen_061918.pdf
18. M.G. Aartsen et al., Phys. Rev. Lett. **115**, 081102 (2015)
19. T.K. Gaisser, E. Resconi, O. Schulz, Phys. Rev. D **79**, 043009 (2009)
20. M.G. Aartsen et al., Phys. Rev. Lett. **111**, 021103 (2013)
21. R. Abbasi et al., Phys. Rev. D. **104**, 022002 (2021)
22. M.G. Aartsen, et al. arXiv:1710.01191 (2017)
23. M.G. Aartsen et al., Eur. Phys. J. C **79**(3), 234 (2019)
24. J. Stettner, PoS **ICRC2019**, 1017 (2020)
25. M.G. Aartsen et al., Phys. Rev. Lett. **125**, 121104 (2020)

26. M.G. Aartsen, et al. [IceCube], Astrophys. J. **809**(1), 98 (2015). arXiv:1507.03991 [astro-ph.HE]
27. T. Glüsenkamp, EPJ Web Conf. **121**, 05006 (2016)
28. J. Stachurska, EPJ Web Conf. **207**, 02005 (2019)
29. M.G. Aartsen, et al., Science, **361**, eaat1378 (2018)
30. M.G. Aartsen et al., Science **361**, 147–151 (2018)
31. Robert Stein et al., Nat. Astron. **5**(5), 510–518 (2021)
32. M.G. Aartsen, et al. [IceCube], Nature **591**(7849), 220–224 (2021) [erratum: Nature **592**(7855), E11 (2021)]
33. J.A. Aguilar et al., Astropart. Phys. **23**, 131 (2005)
34. S. Adrian-Martinez et al., Astrophys. J. **786**, L5 (2014)
35. S. Adrian-Martinez et al., Astrophys. J. **760**, 53 (2012)
36. A. Albert et al., Phys. Rev. D **96**, 062001 (2017)
37. A. Albert et al., Astrophys. J. **892**, 92 (2020)
38. A. Albert et al., Astrophys. J. Lett. **850**, 35 (2017)
39. A. Albert et al., Astrophys. J. **870**, 134 (2019)
40. A. Albert et al., Astrophys. J. Lett. **863**, 30 (2018)
41. S. Adrián-Martínez et al., JCAP **12**, 014 (2015)
42. M.G. Aartsen et al., J. Phys. G **48**(6), 060501 (2021)
43. M.G. Aartsen et al., arXiv:1510.05228 (2015)
44. M.G. Aartsen et al., J. Phys. G **44**, 054006 (2017)
45. S. Adrian-Martinez et al., J. Phys. G **43**(8), 084001 (2016)
46. A.D. Avrorin et al. Phys. Part. Nucl. **46**(2), 211–221 (2015)
47. M. Agostini, M. Böhmer, J. Bosma et al., Nat. Astron. **4**(10), 913–915 (2020)
48. G.A. Askaryan, JETP **14**, 2 (1962); **48**, 988 (1965)
49. M.A. Markov et al., Nucl. Inst. Methods A **248**, 242 (1986)
50. E. Zas, F. Halzen, T. Stanev, Phys. Rev. D. **45**, 1 (1992)
51. D. Saltzberg et al., Phys. Rev. Lett. **86**, 2802–2805 (2001)
52. P.W. Gorham et al., Phys. Rev. D **72**, 023002 (2005)
53. P.W. Gorham et al., Phys. Rev. Lett. **99**, 171101 (2007)
54. P.W. Gorham et al., Phys. Rev. Lett. **117**, 071101 (2016)
55. P.W. Gorham et al., Phys. Rev. D. **98**, 022001 (2018)
56. P.W. Gorham et al., Phys. Rev. Lett. **121**, 161102 (2018)
57. M.G. Aartsen et al., Astrophys. J. **892**, 53 (2020)
58. P. Allison et al., Astropart. Phys. **35**, 457–477 (2012)
59. P. Allison et al., Astropart. Phys. **70**, 62 (2015)
60. S.W. Barwick et al., Astropart. Phys. **70**, 12 (2015)
61. H.R. Allan, J.K. Jones, Nature **212**, 129 (1964)
62. P.W. Gorham, C.L. Hebert, K.M. Liewer et al., Phys. Rev. Lett. **93**, 041101 (2004)
63. S. Buitink et al., AIP Conf. Proc. **1535**, 27 (2013)
64. T. Huege et al., PoS, **ICRC2015**, 309
65. The Pierre Auger Collaboration, in *Advances in High Energy Physics*, **708680** (2013)
66. A. Aab et al., JCAP **11**, 4 (2019)
67. The LIGO Scientific Collaboration, Class. Quantum Grav. **32**, 074001 (2015)
68. F. Acernese et al., Class. Quantum Grav. **32**, 024001 (2015)
69. G.A. Askaryan et al., Nucl. Instr. and Meth. A **164**, 267 (1979)
70. J.G. Learned, Phys. Rev. D. **19**, 3293 (1979)
71. R. Lahmann, G. Anton et al., Astropart. Phys., **65** (2015), 69
72. R. Lahmann, Nucl. Part. Phys. Proc. **273–275**, 406–413 (2016)

Printed in the United States
by Baker & Taylor Publisher Services